HUNGARIAN EAGLES

A Magyar Királyi Honvéd Légierő 1920-1945

Gyula Sárhidai György Punka Viktor Kozlik

HIKOKI PUBLICATIONS

Captions to illustrations opposite, from top:

Reggiane Re 2000 *Héja* I, 'V.401', was one of the earliest Italian-built examples to be supplied to the MKHL. Seen here in August 1940, during the period of mobilisation against Romania, the aircraft later served with the *Repülő Akadámia* – Air Academy – as a trainer. In 1942 it was re-engined with an Hungarian WM-14B motor

Reggiane Re 2000 *Héja* I, 'V.4+12', is seen as it appeared on the Eastern Front in 1942 when it was flown by *Hadnagy* – Second Lieutenant – József Béjczy of the *2/1. Vadászszázad* – 2/1. Fighter Squadron. Ultimately, the aircraft was deliberately destroyed by the Hungarians on the 13th January 1943 in the face of the massive Soviet onslaught which destroyed most of the Hungarian 2nd Army

MÁVAG *Héja-M*, 'V.4+87' was Hungarian-built and was used as a fighter-trainer, seen here at Szolnok in 1943

MÁVAG *Héja-M*, 'V.5+57' was the 87th Hungarian-built example of the type and, like many others, served for advanced training at Szolnok and Mátyásföld during 1944

First published in Great Britain in 1996 by
Hikoki Publications
16 Newport Road, Aldershot, Hants, GU 12 4PB

ISBN 0 9519899 1 X

Text by Gyula Sárhidai
Colour drawings by Viktor Kozlik
Tone drawings by György Punka
Line drawings by László Benczur
Photographs are credited as shown
Edited by Barry Ketley
Design by Hikoki Publications
Printed in Great Britain by Martins the Printers, Berwick on Tweed
Distributed by Books International, Farnborough, Hants

Publisher's note:
This publication would not have been possible without the help and encouragement of a number of people, both in Hungary and England, over a long period of time.
Grateful thanks are due to
Mr László Ujházy, Mr László Hingyi, Dr Éva Haraszti-Taylor,
Mr Gyula Sárhidai and Mrs Kinuko Ketley

The structure of the Hungarian language is such that many names, particularly those of military formations, can be written as single or multiple words, with or without capital initials. For consistency, all Hungarian terms in this book are given with a capital initial and are italicised. Additionally, in Hungarian, names give precedence to the family name. Here, all names are given in the English order, i.e. family name last

Captions to front cover:
Messerschmitt Bf 109G-14/AS, Yellow 15, possibly Werknummer 511006, of the *101/III Vadászezred, III Osztály* ie. 101st Fighter Regiment, 3rd Group, seen at Raffelding in April 1945. One of the last survivors of the *Magyar Királyi Honvéd Légierő*, it was probably destroyed on the ground, along with several others, as a last act of defiance by the Hungarian airmen in May as American tanks rolled in. Note the yellow markings adopted by the Luftwaffe, of which the 101/III was a component, in early March 1945
Left hand emblem: The group commander's marking of the *1/II Vadászosztály*, carried on a Fiat CR 32
Centre emblem: The unit badge of the *101. Vadászezred* – the *Pumas*
Right hand emblem: The group commander's marking of the *1/I Vadászosztály*, carried on a Fiat CR 32

Caption to inside rear cover:
Heinkel He 112B-0, V.303, was one of the three examples purchased by the Hungarians for trials. it was actually preferred over the Bf 109 by the test pilots, but no more were to see service with the MKHL. It is depicted here as it appeared in 1940, probably when it was still undergoing trials with the *Repülő Kisérleti Intézetnel - RKI* – Air Experimental Institute, the Hungarian equivalent to the Royal Aircraft Establishment

Caption to picture opposite:
Junkers Ju 88D, F.915, of the *1/1. Távolfelderítő repülőszázad*, seen at Kharkov in 1943. The first three digits of the fuselage code are most unusual, being red, outlined black

FOUNDATIONS

The beginnings of the independent Hungarian air force

Prior to 1914, the numerous ethnic groups which formed the Austro-Hungarian Empire – the Dual Monarchy – had required that the Army be organised into three distinct components. These were the Landwehr, composed mostly of German-speaking troops; the Honvédség, composed mostly of Hungarians, and the Common Army, which mingled the dozen or so different nationalities. All owed direct allegiance to the Emperor.

During the First World War the air squadrons of the Common military forces – K.u.K. Heer – of the Austro-Hungarian Monarchy were not independent at all. As 'Air Squadrons' they were part of the army, consequently the squadrons, like the army, had both Austrian and Hungarian personnel. Small numbers of other nationalities also served as aircrew members. It says much for the abilities of these airmen that in the successor states formed after 1918, some of the ex-officers of the Common air squadrons became the heads of their own independent Air Forces. This became the case in both Poland and Czechoslovakia.

In November 1918, an Hungarian Republic was proclaimed and somehow, from amongst the turmoil of the retreating and disintegrating Common army, an Hungarian Air Force emerged. This mainly used aircraft which were left in the home factories and in the training centres because very few machines came back from the front lines. Hungary was quite well off for factories, as the main production capacity of the Dual Monarchy had been established in that country. Besides Budapest, the capital, the most important aircraft factory was established at Albertfalva, the so-called UFAG – *Ungarische Flugzeugwerke AG*. Aircraft were also built at Mátyásföld in the Hungarian MÁG – *Magyar Általános Gépgyár* – factory, and at Aszód, in the Lloyd – *Ungarische Lloyd Flugzeug und Motorenfabrik* – works. In Arad, the so-called 'MARTA' factory produced engines, mostly Daimler types, whereas Hiero engines were produced in Budapest in the MÁG factory. Between 1914-1918 the total production of the three Hungarian factories was some 2,029 aircraft and 1,146 engines.

During the tenure of the Károlyi Government, which only lasted five months, about six embryonic air squadrons came into existence. In the prevailing conditions they were incapable of fighting and were unable to take part in the defence of the country against the invading Romanian and Czechoslovak armies.

When the short-lived Hungarian Soviet Republic was founded after 21st March, 1919, the air squadrons were named the Hungarian Red Air Force, under the control of the 37th Department of the Defence Ministry. Organised into eight air squadrons, numbered 1 to 8, each volunteer army division had a squadron, whereas 9. Squadron operated marine aircraft. The whole force fragmented after 1st August, 1919, as there was neither ground support and spares, nor operational bases. Nearly half the country fell under the control of the Romanian army in August and September and what was left of the air force escaped westwards to the trans-Danubian part of the country. One squadron was left in Szeged under the control of the French.

In the ranks of the army

In May 1919 the military and gendarmerie forces in and around Szeged were integrated and became the basis of the new army of the provisional Government of Count Gyula Károlyi. On 5th June 1919, Vice-Admiral Miklós Horthy of Nagybanya arrived in Szeged, and became the Minister of War in the Károlyi Cabinet. On 6th June a decree was issued for the formation of a national Hungarian army and at the end of the month a Supreme High Command was created.

By the end of July this army already had many aircraft. More and more arrived from areas of the Hungarian Soviet Republic. On 1st August, Red army resistance on the front line at Tisza on the plains east of Budapest collapsed, although elements of the force escaped to the trans-Danubian area and were saved. The Romanian army advanced quickly over the great Hungarian Plain and on August 6th they occupied Budapest. After a couple of days they progressed to the trans-Danubian area and on its northern part towards the west they occupied Győr too. As a result of this, they began to rob the whole country, taking away every movable piece of military equipment as booty. The MARTA factory at Arad had already been occupied in November 1918 and soon thereafter all three of the factories and airfields around Budapest were also seized. In September, the Romanian forces dismantled and took away the entire gun-factory at Győr, and the same fate was to befall most parts of the other aircraft factories.

The Allied Peace Committee organised the Hungarian Allied Military Mission and by order of the High Consultative Body the Romanians had to leave the trans-Danubian area on 11th October, 1919. On 13th November they were obliged to leave Budapest. Following this the Romanian forces gradually evacuated the country. They had repaired the important railway bridge at Szolnok by 28th August 1919 and over this bridge, and that at Csongrád, they

Above: *A poor quality, but extremely rare picture of a Hansa-Brandenburg C.I in the markings of the short-lived Hungarian Soviet Republic. Note the camouflage on the tailplane, reminiscent of the later Luftwaffe 'splinter' camouflage. Summer 1919 (Jávor)*

Left: *How it began – and ended – in 1919. Phönix C.I HP.2, ex-121.82 of the Austro-Hungarian air arm after coming to grief on one of the soft Hungarian airfields.. Engine is a 230 hp Hiero (Ketley)*

Left: *Udet U12, 'a', the very first of the type in Hungarian service. The condition of the aircraft suggests that the picture was taken soon after delivery in 1925, probably at the REGVI – Government Pilot School – at Szombathely. Note the varnished ply fuselage and the small pennant on the lower wing trailing edge (Ketley)*

removed from the country 1,292 locomotives and 54,130 wagons full of looted and valuable material by the time of their final departure in April 1920.

The National Army forces coming from the trans-Danubian area marched into Budapest on 13 November, 1919 and regained their supremacy over the Capital.

From the beginning of August in the re-christened War Ministry, the so-called '37 Air Department' was created to deal with aviation problems, but had only a grand total of ten elderly military aircraft available.

In Szeged, training began in August 1920 on Brandenburg, Phönix, Fokker D.VII and UFAG C.I aircraft of the old 4th air squadron. In November three D.VII, one D.VI, three UFAG C.I and one Brandenburg 169-Series C.I supported the march back into Budapest.

There were three bases under the supervision of the National Army: Szombathely, Siófok and Szeged, and later Budapest. Consequently three air squadrons were organised and were active until February 1920.

The Trianon Peace Treaty

On 15th January 1920, the High Command of the Allied Powers gave the Hungarian delegation in Paris the conditions for peace.

By February it became known that the victorious powers would not only prohibit military aviation by Germany but by Hungary as well. To circumvent this, the Hungarians made arrangements whereby the command of the Hungarian National Army air force aircraft on strength would be taken over by the Department of the Airline of the Trade Ministry. They organised the Hungarian airline company – MAEFORT - *Magyar Aeroforgalnis R.T.* – and the military aircraft were inducted into this company as passenger machines. All armament was removed and every aircraft received a civilian registration consisting of a black 'H' prefix and four letters on a white background.

The peace conditions for Hungary reduced the area of the country from 282,000 square kilometres to 93,030 square kilometres and the population from 18 million to 9.5 million. Thus 3,263,000 Hungarians became citizens of foreign countries under hostile administrations. The main military decrees of the peace conditions were the following:

General military service was forbidden, and the army was to consist of no more than 35,000 men. Training of officers was restricted; it also curbed the formation of gendarmerie, police and frontier guard units; confiscation of all the war materials which could be found and all anti-aircraft guns; forbade the existence or use of guns larger than 105 mm calibre, tanks and all heavy armaments; forbade military flights and ballooning and their production; confiscated all the warships except six patrol boats; forbade the production of guns and gunpowder and their exports. Apart from the framework given for the armed forces, arms production or purchase was made generally impossible.

These instructions were intended to ensure that

Left: *H-MAAB, the sole example of the unsuccessful Oravecz I training parasols, seen at Mátyásföld in spring 1924 (Sárhidai)*

Below: *An AVIS III fighter prototype with an experimental propeller spinner. It was not accepted for service (Sárhidai)*

the security of the two newly created successor states, Czechoslovakia and the SHS Kingdom (from 1927, Yugoslavia), should be guaranteed, as well as ensuring decisive Allied military supremacy. The military potential of the Romanian kingdom doubled compared to her former size.

As the Hungarian Government could do little else, her representatives signed the Trianon Peace Treaty on 4th June, 1920. This treaty lodged the powers to work out the military restrictions in the Council of the Ambassadors. The exclusion clauses became valid only after the ratification of the treaty, consequently there was a year left to save what could be saved. It was evident from the very beginning that the nation would not accept these dictates, if they could be evaded. Hungarian foreign policy made no secret of the aim to regain the lost territory. For the moment they signed, because there was no other alternative, but the aim was to make it invalid as soon as possible. As a result, a part of the 3rd squadron flew under the MAEFORT banner in the early part of 1921. The better aeroplanes were dismantled and hidden on the properties of reliable people – in estate buildings, granaries, barns etc. The same thing happened with the engines. This stratagem was code-named "E-action" and was more or less successful.

Ratification of the peace treaty by the Hungarian Parliament took place on 26th July, 1921, and thus the prohibitions on flight and the building of aeroplanes were enacted. Originally the order should have lasted for six months, but it was made valid until the end of 1922, because the task could not be fulfilled earlier.

Under the auspices of the French Colonel,

Barres, the Aviation Supervisory Committee ordered the destruction of 119 aircraft, 77 engines, more than 200 propellers, and even the sophisticated machine tools of the aircraft factories, although the peace treaty did not include orders for this. The MAEFORT company, because of its covering character, was liquidated on 31st December 1921.

The Aviation Supervisory Committee left in the spring of 1922, but left behind as a watchdog the Inter-Allied Supervising Committee. On 26th February 1922, the prohibition on flying was supposed to be discontinued, but it was extended for one more year. During this time four Hungarian aircraft factories ceased to exist. These were MARE and MÁG in Budapest, Lloyd in Aszód, and Ganz-FIAT, also in Budapest. Some of them merged with other firms. In 1924 their original activities were given up.

The Secret Air Force

1922-1927

Department II of the Trade Ministry still existed and was transformed in August 1921 into Department XI (Air). The leader of the II, then the XI Department, was *Ezredes* – Colonel – István Petróczy, an experienced pilot from the First World War. In 1921 he had already advised the use of the German practice, i.e. the creation of civil airlines and the building up the airline network. These were not forbidden activities and could be used for basic and secondary training, for disguising and retaining the personnel who could become military crews.

With this in mind, systematic training of pilots began in 1922 in Szeged under the command of *Százados* – Captain – László Háry. In 1924, under the command of *Alezredes* – Lieutenant-Colonel – Waldemár Kenese, an aircraft school for pilot training began activities in Szombathely, solely as a civilian institution.

On 10th April, 1924, with the permission of the government, the Air Office – *Légügyi Hivatal - LÜH* – was established inside the Trade Ministry. Its first chief was *Alezredes* – Lieutenant-Colonel – Károly Vassel, of the headquarters staff – *vezérkari szolgálatot teljesitő -v.k.szt.* Under his leadership, a countrywide system of airports was built and thereby saved most of the staff for better days under the guise of weather forecasting, airmail and passenger aircraft duties. By December 1927 there was a total of aviation 548 personnel.

From 1922 the European political situation changed enormously. Italy broke away from her former allied sympathies when Benito Mussolini gained power, thus Hungary's political isolation ended. In March 1927, the Council of the Ambassadors declared that they had terminated the activities of the Inter-Allied Military Supervisory Committee and lodged the power in the hands of the Hungarian Government. At nearly the same time the four allied powers, France, Great Britain, Italy and Japan, signed an air agreement with Hungary. Essentially they put an end to those sanctions, known as the London Ultimatum, which had severely effected the air industry and the whole of civilian flying activities.

By this time there were hardly any resources left in the country suitable for aircraft production; there was only a facility at Székesfehérvár-Sóstó which had been transferred from Szeged. This was more akin to an aircraft repair workshop, producing wooden aircraft, but unable to manufacture engines.

Hopes to produce modern aircraft depended upon factories, therefore the authorities investigated the plant of the Hungarian General Engineering Works – MÁG – at Mátyásföld, which existed near the airfield but produced other items. Some thought was given to building a glider and engine section at the Weiss Manfred Company in Csepel, an idea which was realised in 1928.

Above: *Fiat CR 20bis fighters, still in Italian markings, shortly after their arrival at Szombathely in 1932. These were the first real fighters to serve with the MKHL (Sárhidai)*

Left: *Heinkel HD 22, H-MBEH, at Szeged in July 1937. Thirty were built under licence by WM and served as general-purpose trainers and mailplanes from 1927-1933 (Sárhidai)*

Left: *Fiat CR 32 fighters of the 1/2. 'Ludas Matyi' Vadászszázad, Börgönd, 1938. Nearest is V-169 (Sárhidai)*

During the period in which military aviation could only be carried on by subterfuge, the authorities had tried to maintain Hungarian aircraft production but were unsuccessful. In 1923 István Petróczy resigned and retired because of the failure of the *'Szebeny'* training aircraft and of the so-called *'Fejes'* engine. The *'Oravecz'* I and II type training aircraft were equally unsuccessful. Thus, instead of encouraging the production of national designs, Károly Vassel accepted the old K.u.K officers' advice and backed the import of German and Austrian aircraft or their production by licence.

The result of this policy was the gradual acquisition of many old and used British and Italian aircraft and, later, German ones. In this fashion the Udet U12/12a and the Heinkel HD 22 arrived in Hungary. Ernst Udet later made a good deal for himself by selling the production rights to out-of-date aircraft under licence to the WM factory in 1927, which commenced to produce them for years afterwards. As a result the ranks of the Hungarian air force became regrettably full of old type aircraft which should have been placed in museums. The effects of this policy were felt with a vengeance in the 1940's.

In the 'hiding' period the secret air force began to re-use two Brandenburg C.I spotters and four Fokker D.VII fighters, fitted with newly-purchased engines. Besides these, there were eight reconstructed Brandenburg B.I (KBL) aircraft, which were used only for training, exercises and courier tasks. These types were not suitable for military purposes, and had no military equipment.

To summarise, during the eight years between 1920-1927, there was a great gap in the continuity of the Hungarian air force; a whole generation failed to receive proper training and those pilots who took part in the First World War could not continue their profession. Both the knowledge and the practice remained at a standstill in Hungary while aviation in the rest of the world developed at great speed.

The Secret Air Force: 1928-1938

The situation of Hungary with regard to foreign relations gradually improved, for besides co-operation with Italy, when radical Germany appeared on the scene, European relationships changed greatly. Substantial German-Soviet co-operation, lasting until 1934, heavily burdened the successor states to the Austro-Hungarian Empire, which were mainly formed with French assistance.

Hungarian political and military leadership at the highest levels was naturally very sensitive towards every change and tried to use them for strengthening the *Magyar Királyi Honvédség* – Royal Hungarian Army. These aims were hindered by the clauses of the peace treaty and the poverty of the nation, which had been made even worse by the looting of the former years and by the great economic crisis in the years 1929-32.

Károly Csáky, the Minister for Defence at the December 1927 Privy Council meeting, presented a draft plan under which the best part of the Army should be saved for "better times". The Regent, Miklós Horthy, received many memoranda referring to the defence forces of the country. Count István Bethlen, during his visit to Rome in April 1928, asked for 400 aircraft from Mussolini. The wish was to raise the number of air force personnel to 1,189 by the year 1929.

On 16th December, 1928, *Altábornagy* – Lieutenant General – Vassel was appointed to a post which did not exist on paper: Inspector of the Air Force. Leadership of the LÜH fell on Doctor György Rákosi who served on the General Staff as a Colonel – *Vezékariezredes.* By these means they succeeded in organising the Testing Institute of Flight Doctors, the School for Aircraft and Engine Construction and the Aerial Weather Forecasting Service.

As the Locarno Treaty in 1925 and the Paris Agreement in 1927 gradually eased the limitations on aircraft production, the MÁG was involved in negotiations with the Dutch Fokker factory in an attempt to discover what types of primarily civil aircraft they had available and which could be of use to the concealed Hungarian air force.

Even during the period of complete concealment there were nine air squadrons forming three air group cadres in the MKHL – *Magyar Királyi Honvéd Légierő* – Royal Hungarian National Air Force, then under the control of the air department. The following were the cover organisations:

The National Aircraft-Pilot School at Szombathely – *REGVI*; The Aero Club at Debrecen – *DAC*; the Matra Aero Club – *MAC* – in Miskolc; The Aero Club of Somogy County – *SAC*; the Technological University Sport Flying Society – *MSrE* – at Érd; the Budapest Sport Union – *BSE* – at Mátyásföld and the Experimental Flight Group – *Repülő Kísérleti Csoport* – at Székesfehérvár.

Fokker offered for sale the two-seat C.V. general-purpose type aircraft which was capable of many duties, and also the F.VIII twin-engined passenger aircraft with 420 hp Bristol Jupiter radial engines. Of course, neither type was fitted with military equipment. These were to be the training aircraft for reconnaissance, for fighters and for bombers. A second type of training aircraft was bought from the Germans, namely the Heinkel HD 22 mailplane, which had the facility of being able to use old engines.

With the need to organise home production, the MÁG and Weiss Manfred Rt. factory were given commissions to create aircraft components. At the beginning of 1928, an engine factory was built at Csepel, and here the famous Bristol Jupiter class of engines was produced. Later, most of the Hungarian-made engines came from here.

From 1930, those who belonged to the staff of the LÜH received uniforms complete with aviation type dress daggers. In a country where there was hardly any airline network, an enormous meteorological web was built up as the concealed aircraft squadrons appeared as the '1st and 2nd aerial photographic sections, weather forecast and aircraft passenger groups' which sounded convincingly innocent.

Following the granting of introductory trade credits, Italy sent Fiat CR 20 fighters and Caproni

Above: *A Fiat CR 42 of the 1/4. 'Szent György' Vadászszázad on display in the City Park in Budapest in spring 1941 (Sárhidai)*

Right: *Fiat CR 32 fighters of the 1/3. 'Puma' Vadászszázad undergoing maintenance in the summer of 1941. Note the camouflaged propeller blades and the small unit emblem under the cockpit (Punka)*

Ca 101 bombers, thereby strengthening the Hungarian air force. This disturbing fact was known to neighbouring countries, but because of the political situation they were unable to do anything about it.

With the development of home aircraft production in mind, Dr. György Rákosi gave orders for the construction of the AVIS experimental fighters. Not only this, orders were placed the year after, 1930, for Italian aircraft imports. Orders for German aircraft imports followed in 1935. Consequently, 76 CR 32 fighters soon arrived. Thereafter 66 Junkers Ju 86K-2 bombers, 46 Heinkel He 46E-2, 5 He 45 and 18 He 70K reconnaissance planes were delivered to Hungary. Including training machines, German imports amounted to 190 aircraft.

From 1933-34 the reconnaissance biplanes carried military equipment, mainly twin 8 mm Gebauer machine-guns for the pilot and an M.34 type 8 mm Gebauer observer's gun, but the national markings still remained the civilian 'H' prefix and four letter sequence.

The government aimed for total liberalisation of the armaments industry and parity of equipment with the neighbouring nations who were signatories to the 'Little Entente' with France, i.e. Czechoslovakia, Romania and Yugoslavia. Naturally, the air force was aware of this and gave considerable attention to how a potential enemy might behave.

In 1934, Czechoslovakia had 546, Yugoslavia 470 and Romania 773 military aircraft – not to mention the Navy – and the Hungarians had only 200.

In 1937, in order to bolster home production even more, the Hungarian Wagon and Engine Factory – *M.W.G. Rt.* – in Győr also organised their own aerial department for the repair of different aircraft types.

After 1933, foreign relationships in Europe saw a fundamental change. In March 1935 Germany introduced conscription and denounced the clauses of the Versailles Treaty, then within a short time entered into a naval agreement with Great Britain. The Rome-Berlin axis was hurriedly set up and its power became decisive in Europe, by defining the two power groups of the Axis and the Entente and their satellites in Eastern Europe. Poland never backed this delineation, thus Hungary's situation improved, with consequent effects upon the air force. In 1935-36 there were three fighter, three reconnaissance and eight bomber squadrons. In March 1936 the LÜH suggested that two flight brigades should be created with four regiments, which altogether would have formed 26 squadrons which would have meant that the MHKL would have expanded to 12,500 men. This plan was realised only in 1942, because there were many intervening difficulties. By 1937 the air force had only 25 feeble squadrons and 255 military aircraft, although they finally had both regular and reserve personnel and the framework for a stronger force in the future.

Above left: *Hadnagy – Second Lieutenant – Mátyás Pirithy standing in front of Fiat CR 32, V-163, of the 1/6. 'Kör Ász' Vadászszázad (Pirithy)*

Below left: *Fiat CR 42, V-249, of the 5/1. 'Bagoly' Vadászszázad (Sárhidai)*

IN THE OPEN

The Air Force as a branch of the military services

1938-1940

The year 1938 brought a decisive change in the Hungarian Royal Army's situation and development. Relying on the German-Italian alliance, backed by the Polish and Austrian governments, the Hungarian Government revealed her rearmament programme. On the 5th March 1938, in Győr, the Prime Minister, Kálmán Darányi, announced his 1 billion pengőes (the currency of the period) rearmament programme which, disregarding the political and financial barriers, aimed to accelerate the rebuilding of the national defence systems within 3-5 years. This battle-order plan, which allowed for an air division, had already been accepted under the name 'Huba' by the *Koronatanács* – Privy Council – on 22nd February. In the first year there were to be 55 million pengőes available for the procurement of new aircraft.

Recognition by the 'Little Entente' states that from their point of view the international situation had deteriorated, led them to sign an equal rearmament pact with Hungary, in Bled on the 22nd August 1938. From this time on rearmament was legal, the only problem was the impoverished state of the nation. Until the outbreak of war the air force struggled to overcome 10-12 years of technological backwardness but was only partially successful.

On 1st September, 1938, shortly before the Munich Crisis, the MKHL introduced the red-white-green chevron national military markings. Letter and number combinations identifying units

Above: *Three Heinkel He 46E-2Un spotters from the 'Hollo' – Raven – Közelfelderítő század – tactical reconnaissance squadron in 1939-1940. F-307 and F-310 nearest to camera, both carrying the unit emblem on the fuselage (Sárhidai)*

Below: *Closeup of the emblem of the 4/1. 'Uz Bence' Bombazószázad of the 4. Bombazóosztály on the nose of a Caproni Ca 135bis. Debrecen, summer 1941 (Sárhidai)*

Left: *Observer's in-flight view of an Heinkel He 46E-2Un of the II. 'Lucifer' Tactical Reconnaissance Squadron based at Székesfehérvár during the mobilisation against Slovakia in 1939. Modellers should note the crudely applied camouflage paint on the wing struts of the camera aircraft (Ketley)*

Below: *A Szeged-based Bücker Bü 131, I-157, in use as a courier sometime in 1939. Note the very light fuselage finish and what appears to be a red? outline to the metal engine cowling panels. Owners of the ladybird badge are unidentified. A WM 16 'Budapest' light bomber, B-205, probably from the 2/4. 'Vörös Ördög' light bomber squadron sits in the background (Sárhidai)*

and aircraft, and specific paint finishes were also introduced. At the same time the air force was reorganised after long general staff discussions. *Ezredes* – Colonel – Ferenc Feketehalmi Czeydner of the General Staff became the aerial section head; *Vezérőrnagy* – Major-General – Waldemar Kenese became inspector of the air force; *Vezérkari ezredes* – HQ Staff Colonel – Ferenc Szentnémedy, Chief of the General Staff, and *Ezredes* László Háry became head of the air force proper. Vice-Admiral Horthy, the Regent, subsequently ordered that after the regulation of the central management the *Magyar Királyi Honvéd Légierő* should become an independent service with effect from 1st January 1939. In the interests of more officer training, the Royal Hungarian Miklós Horthy Military Air Force Academy in Kassa was instituted from 31st March 1939. With the rising numbers of personnel there was a consequent need for more training resources. In 1939, with the help of the Italians, 100 officer candidates and 100 NCO candidates went south to Taranto to undergo Italian pilot training – *DRT* – courses. Later, in 1940, another group of 200 went there, but no more were sent after that because the fundamental aim became to acquire more training aircraft and train the pilots at home following the German pattern.

The major obstacle to development was mainly the lack of aircraft, as home production was still unresolved. Only the WM company, the MWG and the MÁVAG aircraft sections were working well, but apart from the WM-21 *'Sólyom'* – Falcon – observation aircraft, they could not mass produce anything. Between them they built 128 aircraft between 1939 and 1942, but this had little effect. The aircraft repair plants in Debrecen and Székesfehérvár were adequate for their tasks, but were unable to develop production. The fact was that the factories of the nation were still not equipped for the production of metal military aircraft and there was no supporting instrument and component industry. In 1939 the MKHL had 6,075 men, but instead of the intended establishment of 304 military aircraft they had only 252 at their disposal, thus the operational squadrons were not fully up to strength either. Despite the best intentions, hopes of importing more modern machines came to nothing, as Germany was unwilling to sell such aircraft. Italy, on the other hand, only offered mainly out of date types, as she could not produce enough for her own needs. All the same, because of the urgent need, in the autumn of 1939, the Hungarians ordered 50 Fiat CR 42 fighters. An order for 70 Reggiane Re 2000 *Falco* I fighters followed at the end of the year and another at the beginning of 1940 for 36 Caproni Ca 135 bis bombers as well.

Conflict with Slovakia

When the southernmost portion of the previously Czech province of Slovakia, with its large ethnic Magyar population, was annexed by Hungary on 2nd November 1938, the military districts of the country were extended but not the air force, because there were no aviation facilities in the returned areas. At the time 11,927 square kilometres of land with a population of 1.03 million people came back to the nation. The city of Kassa was a potential home for the Horthy Military Aircraft Academy, which was actually destined never to be built. Instead, the Academy was formed in buildings on a site not very far from the airfield there.

On 14th March 1939, the Slovak Republic was announced, and on the night of the 14th-15th, in negotiations held in Berlin, Adolf Hitler, as *Führer* and Chancellor of Germany forced the Czech leadership to recognize the Bohemia-Moravian protectorate. The troops of the Wehrmacht marched into Prague and occupied the whole area on the 15th of March. Thus one state from the Little Entente disappeared, representing one third of the hostile states around Hungary. The military forces of the new Slovak nation did not threaten Hungary, but the feelings of hostility remained.

On 15th March 1939, Hungarian troops began their re-entry into Ruthenia, the largely Magyar eastern province of Czechoslovakia, which was already in a state of political turmoil. These troops reached the old frontiers at the Carpathian Mountains on the 24th. In this area the frontiers were still subject to disputes, as the whole area had once belonged to the Hungarian Kingdom. On the orders of the General Staff, the Hungarian troops

Above: *Caproni Ca 101/3m in service as a paratroop transport at Pápá in 1940. The small Hungarian paratroop unit enjoyed a similar relationship to the MKHL as the German paratroop force did to the Luftwaffe (Sárhidai)*

Right: *A Focke-Wulf Fw 58 courier aircraft and 'chauffeur' at Debrecen in August 1940. Reggiane Héja I fighters in the background (Sárhidai)*

Left: Aircrew briefing for men of the II. Közelfelderitő század 'Lucifer' – II. Tactical Reconnaissance Squadron – based at Székesfehérvar in 1938-1939. Man on left wears the shoulder rank insignia of a Főhadnagy – Lieutenant (Jávor)

Below: A Fiat CR 32 fighter of the 1/1. 'Ijasz' Vadászszád – 1/1. Fighter Squadron, ready for action during the brief war between Hungary and Slovakia. The tails of several Focke-Wulf Fw 56 trainers can be seen in the background. Ungvár, 16th March, 1939 (Kovács)

had to reach an easily defendable sector above Sobranc which was to be the new frontier between the two nations. At the same time a clash occurred as the new Slovak Air Force, which had only existed for nine days, began to attack the Hungarian troops.

From 16th March there were two *vadászszázad* – fighter squadrons – in the area, each with nine Fiat CR 32 aircraft; the *1/1. 'Ijász'* squadron in Ungvár and the *1/2. 'Ludas Matyi'* in Miskolc. The *1/3. 'Puma'* fighter squadron was at Csap, but barely operational because of the waterlogged state of the airfield. The *3/4. 'Sárkány'*, the *3/5. 'Hüvelyk Matyi'* and the *3/3. 'Sárga Vihar' Bombázószázad* – bomber squadrons – each with nine Junkers Ju 86K-2 aircraft, went to the airfield outside the city of Debrecen.

In two incidents during the afternoon of 23rd March, Hungarian anti-aircraft gunners shot down three Letov Š-328 and two Avia B-534 aircraft of the Slovak air force. On the morning of the following day, in the area of Ungvár, three Hungarian CR 32 fighters shot down two Slovak Avia B-534 fighters and a third was forced to land. In the south, the Ungvár anti-aircraft defences shot down a Letov Š-328. In the middle of the day, nine CR 32 aircraft of the *1/1. Vadászszázad* – 1/1. Fighter Squadron – shot down four Š-328 and three B-534 aircraft during a fight in the area of Sobranc. At almost the same time, 15 of the 18 available Ju 86K-2 bombers attacked the airfield of Iglo, but with poor results; ten aircraft were damaged on the ground and there was some damage to the edge of the city. There were no Hungarian aircraft losses, but the operation was disorganised and ended poorly, as the 18 Ju 86 aircraft from the bomber wing's second squadron did not arrive from Szombathely in time and therefore the planned joint attack proved to be abortive.

Mobilization – the Transylvanian Crisis

Tension between Hungary and Romania reached a peak in 1940, owing to Romania's failure to respond to Hungarian claims to the disputed area of Transylvania. The Hungarian Air Force was alerted on 27th June when, confronted with an ultimatum from the Soviet Union, the Romanian Government accepted that they would be forced to give up the conquests of 1919 and would evacuate Bessarabia and North Bukovina.

This time it was thought that the Hungarian aims could be solved even at the price of military action, as there was no possibility for the Entente powers to help Romania. On 2nd July the *1. Repülődandár* – 1st Air Brigade – was ordered to transfer to the region of Miskolc, to Tiszalök, Debrecen, Hajdúböszörmény etc. Later the whole air force was moved. The main problem was the imperfect equipment of the Ca 135 bis bombers, with the result that they were unable to be used. The first aircraft to arrive, the Re 2000 fighters, were stationed in Debrecen in order to strengthen the air defences.

At first, because of the negotiations at Turnu-Severin in Romania, the state of alert was cancelled, but later, on the 23rd August, total mobilisation was ordered. The 360,000 soldiers of the deployed 1st, 2nd and 3rd Armies were ready to commence hostilities by the morning of the 28th August, the air force by then having reached the same state.

At that moment Germany intervened as an Hungarian-Romanian war might well have threatened the main source of oil to Germany, therefore Hitler stepped in and forced both Governments to accept German-Italian arbitration. On the 30th August 1940, the second Vienna Award decided that 43,591 square kilometres from Transylvania should be ceded to Hungary, along with a population of 2.18 million. In the meantime, there were several frontier incidents. On the 27th August, a Romanian He 112 fighter attacked an Hungarian Ca 135 bomber between the frontier and Debrecen. The bomber was damaged but was able to land safely, albeit with a dead crewman on board. The second day an Hungarian WM-21 *Sólyom* observation aircraft from Ungvár dropped two bombs on the Romanian airfield of Szatmárnémeti, and was then partly wrecked in landing on Hungarian territory. There were many reconnaissance flights; later still, on 23rd November 1940, a Romanian Fleet 10G training aircraft landed at Fiatfalva followed by an IAR 80 fighter on 16th January 1941 at Kolozskara. Similarly, on 12th October 1940, two Hungarian He 46E-2 aircraft strayed into South Transylvania. Events such as these did little to mend the relationship between the two countries.

Until the end of 1940 the problems of organising the air force increased, as there was no possibility of somehow overcoming the lack of aircraft. The para-troop squadron, with its aircraft, was included in the air force and the decision was taken to develop it into a battalion. By the end of 1940, the Hungarian air force consisted of 5,734 professional and conscript soldiers.

Confronted with a failure to solve the problems of the last fifteen years within two years, the general staff began a witch-hunt, and those whom they regarded as responsible were dismissed. The new plan was to discontinue the independent service which imitated German principles and instead to merge the air force, the anti-aircraft defence and the civilian defence organisation under a central command. Consequently, on the 24th December 1940, *Ezredes* – Colonel – László Háry, was promoted out of his post and into retirement as a titular Major-General. On 1st March, 1941, the new order came into effect and the independence of the air force ended. On the same day András Littay, an infantry general, was nominated as air sub-department head, and Géza Vörös, a general staff colonel, to the head of the Air General Staff. In effect the air force became an air army, which had a sub-department in the War Office.

Above: *A WMA 21 Sólyom of the VII. Közelfelderítő század – VII. Tactical Reconnaissance Squadron – based at Miskolc in 1940 (Sárhidai)*

Left: *A Junkers Ju 86K-2 of the 2/3. 'Buzogány' Bombazószázad – 2/3. Bomber Squadron running up its engines as it prepares for take-off from the field at Börgönd, 1940. Note the unit emblem and open belly hatch (Sárhidai)*

WORLD WAR II

The battles on the Serbian and Soviet Fronts in 1941

The air force was reorganised during the winter of 1940-41 and because of the bad weather there was no flying at all. Between 1938 and 1941, 78.5 million pengőes had been spent on imported German aircraft and another 113 million on Italian machines, but the Hungarian air force had still been unable to import enough. On 1st June 1941, orders were given to establish an air defence army corps to integrate all the forces.

In the summer of 1941, *Altábornagy* – Lieutenant General – Béla Rákosi became the head of the air force, but he, too, was also unable to find a new or better concept for the development of the service, which continued to find demands upon it outstripping its resources.

The next crisis came suddenly, when on 27th March 1941, a military coup overthrew the Cvetkovič government in Belgrade. The German government almost immediately decided that their planned military operation, 'Fall Marita', aimed against Greece, should be expanded against Yugoslavia as well in order to suppress the new anti-German Government. Therefore, besides Italy, the Germans asked Hungary to take part in the operation. The reward would be the return of the lost southern part of Hungary. As over half a million

***Top:** Formation of Junkers Ju 86K-2 bombers of the 2/5. 'Halál' – Death – Bombazószázad – Bomber Squadon – during a training flight from their base at Börgönd in 1940. Red, white and green stripes are clearly visible on the tailplanes, although there seems to be no sign of the wing chevron markings. Note the extended ventral turrets (Sárhidai)*

***Left:** Caproni Ca 135bis bombers of the 4/1. 'Uz Bence' Bombazószázad lined up on the airfield at Debrecen in spring 1941. Nearest is B-508 (Punka)*

Hungarians lived there, the Hungarian government felt unable to deny the request, although they were aware of the political risks of the situation. Had they not acted, there was a danger that Parliament would have overthrown the government. On the 3rd of April, 1941, this responsibility, and the unauthorised passage of the Wehrmacht through Hungary *en route* to Yugoslavia led the Prime Minister, Count Pál Teleki, to commit suicide.

6th April 1941

His successor, László Bárdossy, could not offer an alternative solution, thus the *Koronatanács* – Privy Council – agreed to the partial mobilization of five army corps and the air force. German forces came from Austria, Romania and Bulgaria, a smaller number from trans-Danubia via the river Dráva ready to enter Yugoslavia. As escorts, the Luftwaffe fighter unit II/JG 54 landed at Kecskemét, II/ZG 76 at Szeged and smaller reconnaissance squadrons settled on airfields around Pécs. On the Hungarian side the *1. Repülődandár* – 1st Aviation Brigade – was placed in an attack position on the line of Baranya-Bácska in support of the 3rd Army. This meant that the *1/I., 1/II., 2/I. and 2/II. Vadászosztály* – Fighter Groups; the *4/II. Bombázóosztály* – 4/II. Bomber Group, the *101. Önálló Tavolfelderítő osztály* – 101st. Independent Long-range Reconnaissance Group and the *3/5. Bombázószázad* – 3/5. Bomber Squadron – had to be placed on alert, as well as many airfields.

On 4th April preliminary alert status was ordered; a day later air units were placed at readiness and the frontiers were closed. Meanwhile the German forces occupied the Hungarian airfields. Military operations began towards dawn on 6th April with a concentrated attack by the German-Italian forces in three directions. The Hungarian nation was not yet in a state of war with Yugoslavia, therefore only exploratory flights began.

Yugoslav aircraft flew into Hungarian air space for the first time, reaching as far as Lake Balaton, on 6th April. On the 7th they began bombing attacks against Hungarian stations and airports. In this way bombs fell on Szeged, Kiskunhalas, Siklós, Villány, Pécs, Zalaegerszeg and Körmend. Here the victims were mainly civilians, and the losses of the advancing German forces were few. Severing of the Yugoslav's southern attack route was mainly due to the strength of the 48 German Bf 109E and F fighters from II/JG 54 and the 48 Bf 110D-3 and F Zerstörers from II/ZG 76 based on the airfields of Pécs, Kecskemét and Szeged. The Hungarian *1/3. Vadászszázad* – 1/3. Fighter Squadron – defended Budapest from Mátyásföld and the *2/3. Vadászszázad* – 2/3. Fighter Squadron – protected

Above: *Mechanics working on three Reggiane Héja I fighters of tthe MKHL in August 1940 at Debrecen (Sárhidai)*

Left: *A scene of activity around one of five Savoia-Marchetti S.M. 75 transports used by the MKHL. Apparently nicknamed 'Ernyős' – umbrella – the aircraft were extensively used by Hungarian paratroops. Note the three-tone Hungarian camouflage and yellow wing tips. Pápa, spring 1941 (Sárhidai)*

Kecskemét with their Fiat CR 42 aircraft.

The activities of the Hungarian air forces were directed by István Bánfalvy, the air general staff colonel who commanded the *1. Repülődandár* – 1st Air Brigade. Under his control he had the *1/1, 1/2, 1/3, 1/4, 2/3.* and *2/4. Vadászszázad* – Fighter Squadrons, the *3/5. Önalló Bombázószázad* – 3/5. Independent Bomber Squadron, and the *4/3* and *4/4. Bombázószázad* – 4/3. and 4/4. Bomber Squadrons. These were only active in support of the 3rd Army in order not to interfere with the movements of the German forces. An additional restriction was the fact that there were not enough airfields available in the southern part of the country.

On 7th April, the air defences and the German fighters claimed nine Yugoslav Blenheim and S.M.79 bombers shot down. In Budapest the first air raid warning in the war was on the 6th of April. There were three more in the following days. On 8th April they claimed two more Yugoslav aircraft, but soon afterwards these incursions ceased.

On 12th April two sections from the *4. Bombázóezred* – 4. Bomber Regiment – attacked the Yugoslav defences, but this attack was stopped as counter-orders were received. In the afternoon, the deployment of the first squadron of the parachute battalion from the Veszprém airfield started in four S.M.75 transport aircraft. The lead aircraft, E-101, crashed from a low altitude and caught fire. In this accident, four crewmen from the aircraft, and 19 paratroopers, amongst them *Százados* – Captain – Károly Kelemen, the commander of the air transport squadron, and *Őrnagy* – Major – Árpád Bertalan, the battalion commander, were killed.

In the evening, the three surviving aircraft dropped three officers and 57 men, under the command of *Főhadnagy* – Lieutenant – Zoltán Kiss in the area of Ujverbász, but about 30 km away from the target.

In Yugoslavia the activities of the Hungarian air force ceased on 20th April. Up until this date the *Önalló Távolfelderítő osztály* – Independent Reconnaissance Group's He 70K and He 111P aircraft had made nearly 25 sorties. According to records, the 1st Air Brigade's 60 aircraft flew 600 hours in the air. The losses were: two CR 32, two CR 42, one WM-21 and an S.M.75 destroyed and two CR 32 and one CR 42 damaged. A Blenheim, an S.M.79B and a Dornier Do 17K were captured from the Yugoslavs. The losses were counterbalanced by the arrival of the 70 Reggiane Re 2000 fighters which had been ordered from Italy and which arrived at the end of the year.

There was no time to evaluate the experiences of the attacks against Yugoslavia, before the next crisis arrived. On the 22nd June, 1941, the war between Germany and the Soviet Union started. According to the German war plan, Operation *'Barbarossa'*, Hungary was not to take part in hostile actions. Therefore the defence forces in the Carpathians had to be strengthened by the alerting of the 1st Mountain Brigade, the 8th Mountain Infantry Battalion and the permanent frontier guards manning the defences. The troops of the VIIIth Army Corps were put on alert, followed by the air defences and the civilian defence at a lower level of readiness. The headquarters of the Air Forces and the National Air Defence Command ordered readiness in the morning, the fighter forces being concentrated in the north-east, the other groups not being mobilized. In consequence the following unit deployments took shape:

Unit/Aircraft Used/Base

1/1. *'Dongó'* Squadron Felsőábrány	Fiat CR 32 Defending Ózd-Diósgyőr
1/2. *'Ludas Matyi'* Felsőábrány	Fiat CR 32 Defending Ózd-Diósgyőr
2/4. Air Patrol Squadron Miskolc	Fiat CR 42 Defending Miskolc
2/1. *'Mackó'* Squadron Nyiregyháza	Fiat CR 32 Defending Debrecen
2/2. *'Puma'* Squadron Nyiregyháza	Fiat CR 32 Defending Debrecen
2/3. *'Ricsi'* Squadron Bustyaháza	Fiat CR 42 Defending Huszt
1/3. *'Kör Ász'* Squadron Mátyásföld	Fiat CR 42 Defending Budapest
1/4. *'Szent György'* Squadron Mátyásföld	Fiat CR 42 Defending Budapest
2/1. *'Részei'* Squadron Szolnok	Reggiane 2000 *Héja I* Just forming

According to some accounts the 2/4. Fighter Squadron was also stationed at Bustyaháza.

Right: *The ominous shape of a Junkers Ju 86K-2 bomber of 4/2. 'Isten Kardja' Bombázószázad based at Kecskemét, droning over the Hungarian plain in spring 1941 (Sárhidai)*

Below: *V-413, a Reggiane Re 2000 Héja I fighter of the 2/1. 'Keresztes Pók' – Spider Knight – Vadászszázad, somewhere on the Eastern Front in June 1942.*

A Fateful Decision

On 26th June, 1941, at six minutes past one in the afternoon, the city of Kassa was attacked from a south-easterly direction by three twin-engined bombers which dropped 29 bombs. One did not explode and the nearby village of Enyicke also received a bomb which did not explode either. In the city, where there was no air raid warning, there were heavy losses. 24 houses, three barracks and a post office were destroyed. 32 people were killed, 60 were seriously injured and 220 received lesser injuries. Earlier that day, at 12:40, express train number 1701 was machine-gunned between Tiszaborkut and Biliny by three single-engined aircraft, identified as Soviet Polikarpov I-16 fighters. Three people were killed and six wounded.

The air defences of Kassa, which consisted of three old M. 05/08 guns and four 40 mm Bofors guns, opened fire but achieved nothing. On the following day an incoming signal revealed that the day previously, three unidentified and unmarked aircraft had flown over Ungvár at 10.00 am. The *VIII. Közelfederítő század* – 8th Tactical Reconnaissance – squadron opened fire from the airfield with automatic weapons and the three aircraft left without bombing.

Reports arriving in Budapest were confused and difficult to understand. No-one could have had a clear view about the situation, only one thing was definite: the country had been attacked. In this situation, the military general staff under Henrik Werth, General of the Infantry and Chief of the General Staff, unequivocally decided that the attack had come from the Soviet Union and thus informed the Minister of War, *Táborszernagy* Károly Bartha. This information was given to the Regent, Miklós Horthy and finally to László Bárdossy, the Foreign Secretary. The Regent gave his blessing to a declaration of war and the same afternoon the Cabinet, with two votes against, accepted the text which stated that the government declared a state of war between the two countries as the Soviet attack was directed against Hungarian territory. The Hungarian Ambassador in Moscow passed on this text which, strictly speaking, was not a formal declaration of war. Upon this Great Britain declared war on Hungary.

Investigating the events took some days while the two bombs, each of which weighed 100 kg, were dug out and detonated. From these events, the only remaining evidence is the despatch which enumerated the damage and the narratives of some participants; the primary sources are missing.

Many publications have appeared concerning this question, here we have no space to quote them all. Now, after the war and many years of searching in published Soviet sources, it seems that three Soviet aircraft erroneously flew over the city, failed to find their targets and simply jettisoned their bombs, which were actually intended for somewhere else, although rumours persist that this was a plot engineered by Bárdossy. So far, no-one appears to have the right facts – which aircraft were there, who were the pilots and what were their tasks. Possibly, at some time in the future, when the secret files in the Soviet and Romanian archives are available, the answers will be found.

In December 1941 Hungary foolishly declared war on the United States of America. Thus she became involved in war against practically all the Allies and began her involvement in the battles of the Second World War, which lasted nearly four years. In the first phase retaliatory air raids against Stanislau were ordered on the 27th June and against Strij on the 29th June. For these there were available 25 Ju 86K-2 and Ca 135 bombers and 18 CR 42 fighters. This action was supported by WM-21 aircraft of the *VIII.* and *X. Közelfelderítő század* – Tactical Reconnaissance Squadrons, each of which fielded nine machines.

The first air battle above Hungarian territory occured on the 29th of June when CR 42 aircraft from the *2/3. Vadászszázad* – 2/3. Fighter Squadron – shot down three out of the seven Tupolev SB-2 Soviet bombers which attacked the railway station at Csap. The Hungarian army Mobile Corps was supported by a number of aircraft which consisted of the *3/5. Bombázószázad* – 3/5. Bomber Squadron, *4. Bombázóezred* – 4th. Bomber Regiment and the *2/3.* and *2/4. Vadászszázad* – Fighter Squadrons, plus two tactical reconnaissance squadrons and the *Önálló Tavolfelderítő század* – Independent Long-range Reconnaissance Squadron. During the first ten days the aircraft made 145 sorties and dropped 113 tons of bombs. After 13th July the squadrons began to move to the airfields of the trans-Carpathian territory in order to follow the advancing troops. On 7th August a group of Re 2000's from the *1/2. Vadászszázad* – 1/2. Fighter Squadron – under *Főhadnagy* – Lieutenant – László Gyenes then arrived at the front from Szolnok in order to obtain battlefield experience with the new aircraft.

On 28th August 1941, *Alezredes* – Lieutenant Colonel – Béla Orosz, was recalled and *Alezredes* Sándor Gyiressy was appointed commander of the aviation group.

From 7th July to 28th August the air group made 555 sorties, dropped 40 tons of bombs, claimed 27

Above: *Reggiane Re 2000 Héja I, V-450, of the 5/2. Vadászszázad on display to the public in a park in Budapest, August 1942. Both the Hungarian markings of the period and the machine's similarity to the Soviet I-16 are evident in this view (Sárhidai)*

Right: *Heinkel He 46E-2Un, F-303, of the I. Közelfelderítő század and subject of the 4-view colour drawing in this book, is prepared for action over Yugoslavia. The unit's 'Holló' – Raven – emblem can be seen under the centre cabane struts. Clearly displaying the yellow nose and tail Balkan theatre markings (the yellow fuselage band was only used in Russia), note how the tail chevron can still be made out under the yellow paint (Sárhidai)*

Above: *Pilot and pupil go through their cockpit checks in Arado Ar 96A, G-429, on a fine day on the Don Front in September 1942. Points to note are the various stencil markings, the typical Hungarian 3-tone camouflage and the retractable footsteps (Ketley)*

aerial victories and lost four aircraft and six pilots.

The *I. Közelfelderítő század* – 1st Tactical Reconnaissance Squadron – came back to Mátyásföld on 6th October. From 13th July the squadron had carried out 72 operations, 61 as spotters; flown 115 reconnaissance hours and had made 11 attacks on ground targets, dropping two tons of bombs in the process and did not lose a pilot. The remaining six aircraft from the Re 2000 *'Gyenes'* group returned home after 20 October. From these, two crashed in the Carpathians due to the bad weather. The *1/3. Vadászszázad* – 1/3. Fighter Squadron – returned to Mátyásföld on 26th November. In four and a half months they had made 447 sorties totalling 800 flying hours. They claimed 19 air victories and lost two CR 42 aircraft, János Pettendy being killed and Márton Szőnyi captured.

In December all units of the flying squadrons returned to Hungary, having achieved the following: 1,454 flights over enemy territory, totalling 2192 flying hours. 217 tons of bombs were dropped. 30 Soviet aircraft were claimed shot down; 17 officers and enlisted men 'met hero's deaths'; 4 officers and 2 men were posted missing; 10 officers and 18 men were wounded. Overall aircraft losses totalled 56 machines. 25 long and short-range reconnaissance types; 14 fighters; 11 bombers; 5 courier/liaison types and 1 transport. Of this total, 21 simply disappeared or were totally destroyed. 35 sustained 20-90% damage.

The Eastern Front : 1942

On the 20th January 1942, *Feldmarschall* Wilhelm Keitel, chief of the OKW, arrived in Hungary, following the earlier visit of the German Foreign Secretary, Joachim von Ribbentrop. They declared that the *Führer*, Adolf Hitler, had decided that Hungary was to fully participate in the summer campaign against the Soviet Union. First they demanded that the whole Hungarian Army should take part, but later modified this to only two-thirds. The Hungarian political leadership and the general staff were very much against these demands, but they could not evade the request that the 2nd Hungarian Army should be mobilized and sent to the front. This army consisted of three corps, (III, IV and VII) made up of nine light infantry divisions,

Right: Ground crew cluster round a Junkers Ju 86K-2 of 3/5. ***Bombázószázad 'Hüvelyk Matyi'*** *at Debrecen in summer 1941. How the De Havilland Dragon Rapide in the distance came to be in Hungary is not known (Sárhidai)*

Below: Starboard side view of a WM 21A Sólyom, F-206, of the 4. Közelfederitő század at Ungvár in 1942. The apparently bright colours on the fuselage spine and engine cowlings are actually canvas weather covers (Sárhidai)

each composed of two regiments; the 1st field armoured division and the *1. Repülőcsoport* – 1st Air Division – which was as strong as a regiment and which later became known as the *2. Repülődandár* – 2nd Air Brigade.

The commander of the 2nd Army was *Vezérezedes* – General – Gusztáv Jány, and the chief of the General Staff was *Vezérőrnagy* – Major-General – Gyula Kovács. The 1st Air Group operations staff was formed in Budapest on the 1st of April, 1942 under the command of *Vezékariezredes* – HQ Staff Colonel – Sándor András and *Alezredes* – Lieutenant Colonel – János Németh. The unit was ready for deployment at the front by 21st June. Combat experience with pairs of aircraft led to a switch over to the 12 aircraft squadron organisation after the example of the Luftwaffe.

When the 2nd Army began the march to the Eastern Front, the newly formed *2. Repülődandár* – 2nd Air Brigade – went as well. This was made up partly from two fighter squadrons and partly from squadrons of the peacetime order of battle. On the 12th June the aircraft began operations, and on the 30th July the following units were at the front:

First, on the 12th June, *Alezredes* – Lieutenant-Colonel – Edgár Keksz led the *1/1. Század* – 1/1. Squadron – of the *I. Tavolfelderítő osztály* – 1st Long-range Reconnaissance Group, from Budaörs. Under the command of *Százados* – Captain – János Hollós they had only two old Heinkel He 111P aircraft and an ex-Yugoslav Dornier Do 17K-2. Only later, at the front, did they receive three Do 215B-4's and five He 111P's. Also belonging to the group was a small unit formed from the 1/3. squadron which had no aircraft, but was based at Konotop for testing. After this on the 22nd June, the *3/2. Közelfelderítő század* – 3/2. Tactical Reconnaissance Squadron – under the command of *Főhadnagy* – Lieutenant – Zoltán Trón arrived from Kecskemét with twelve Heinkel He 46E-2 machines. This squadron originally had WM-21 *'Sólyom'* aircraft, but these had very unpleasant flight characteristics and were not fitted with radio, therefore the commander of the squadron successfully managed to get these aircraft exchanged for the old He 46 types. Events afterwards proved the wisdom of this, because after three months operations the squadron had only lost one aircraft and had three aerial victories. The squadron was also subordinated to the *I. Tavolfelderítő osztály* – 1st Long-range Reconnaissance Group's command staff, but this was not a happy decision.

On the 24th June, the *4. Bombázóosztály* – 4th Bomber Group – headquarters staff under the command of *Őrnagy* – Major – István Mocsáry started from Debrecen with the 4/1. Squadron named *'Boszorkány'*, which had been formed by combining it with the *'Uz Bence'* squadron. *Főhadnagy* – Lieutenant – András Inokay commanded the seven Caproni Ca 135 bis bombers.

After some days another four Ca 135 bis followed them.

On the 2nd July, under the command of *Vezerkariőrnagy* – HQ Staff Major – Kálmán Csukás, the staff of the *I. Vadászosztály* – I. Fighter Group, and the *1/1Vadászszázad* – 1/1 Fighter Squadron *'Dongó'*, arrived from Szolnok. Equipped with nine *Héja* I fighters the squadron was led by *Szazados* – Captain – Mátyás Szabó. One aircraft remained at Kolomea as it ground-looped when landing, two others stayed at Kiev for repair. It was in this squadron that the son of Vice-Admiral Horthy, *Főhadnagy* István Horthy, a reservist, volunteered for front line service as a pilot.

The *2/1. Vadászszázad* – 2/1. Fighter Squadron, named *'Keresztes Pók'* and equipped with 11 *Héja* I fighters and two Bücker Bü 131 light aircraft, arrived at the front on the 30th July from Kolozsvár under the command of *Szazados* – Captain – Béla Keresztes.

As a non-combatant unit the Air Brigade also had a courier flight with Ar 96 and Fi 156 aircraft; the 1st air transport squadron with converted Ju 86K's and an Fw 58; the *1/1. repülő-híradó* – air signals – unit, an industrial outfit which delivered urgently needed parts in old Caproni Ca 101 three-engined machines. Later on, the *2. Szállító repülőszázad* – 2nd Air Transport Squadron, which consisted of the Ju 52/3m aircraft of MALERT, supplied the connection with the Air Brigade. This squadron was subordinated to the Hungarian Ministry of War. On their arrival, the air group consisted of 50 combat and 46 other aircraft, with a staff of more than 5,500 personnel. Both the journey and operations were beset with troubles. Certainly, 12 He 46 aircraft had arrived on the 22nd June, but the nine Ca 135 bis bombers which had arrived earlier had lost one of their number when the undercarriage broke on landing and the aircraft became unusable. The *1. Szállító repülőszázad* – 1st Air Transport Squadron – had three Ju 86's and an Fw 58 aircraft, while the industrial unit had three Ca 101's and an Fw 58. On the 27th June an He 46 force landed and the aircraft was damaged. The following day, He 46, F-323, was forced to land, successfully, after being hit by ground fire.

On the 29th of June every available aircraft was used to carry munitions behind the front lines. For this task there were three Ju 86 and three Ca 101 machines. In the evening of the following day, one of the Ca 101's became lost and in a forced landing near Kersen sustained damage. One Ju 86 strayed and had to land near the airfield, during which it nosed-over. Another Ju 86 was fired on in error by German troops; it landed damaged, with one man wounded. A flight of three of *I. Tavolfelderítő osztály* – 1st Long-range Reconnaissance Group's – Ju 86 aircraft were also used for transport duties.

On the 2nd July *Vezérkariőrnagy* – HQ Staff Major – Csukás despatched 12 *Héja* I fighters from

Above: *Ground crew servicing a Junkers Ju 88A-4 belonging to the 4/1. Bombázószázad in May 1943 on an airfield near Kharkov (Sárhidai)*

Right: *A nosed-over Focke-Wulf Fw 58 offers a good view of the Hungarian three-tone camouflage on the upper surfaces. Note the large size of the wing markings. Possibly Szombathely in March 1942 (Sárhidai)*

***Left:** One of the workhorses of the Hungarian tactical reconnaissance squadrons was the Heinkel He 46E-2Un. Two from the I. Közelfederitő század are seen here at Mátyásföld in 1942. Nearest is F-303 (Sárhidai)*

***Below:** Taken from the co-pilot's seat of a Junkers Ju 88A-4, are two other aircraft from the same unit, the 4/1. Bombázószázad, all heading for a target somewhere on the Eastern Front in 1944 (Sárhidai)*

Szolnok to the front, but only 9 arrived at the airfield of Konotop. In the bomber units, of the eight Ca 135 aircraft, only numbers 3, 4 and 5 were in serviceable condition, the remainder suffering from one technical problem after another. On the same day, the Ju 86, B-335, from the transport flight of *I. Tavolfelderítő osztály* and carrying the pilot, *Főhadnagy* – Lieutenant – Stifter, three crewmen and ten wounded troops aboard, was forced to land and in the resulting crash one of the wounded men was killed. *Ezredes* – Colonel – Sándor András, gave orders that an *Héja* I aircraft should be displayed on the nearby German airfields used by Luftflotte 4 and VIII Fliegerkorps fighter units following recognition errors in 1941. Despite this, German pilots in 1942, and later, often failed to correctly recognize the Re 2000 aircraft, believing it to be the Soviet I-16 and consequently opened fire.

On the 3rd July, *I. Vadászosztály* – 1st Fighter Group – was strengthened by another fighter squadron from Szolnok, consisting of 11 *Héja* I's and two Bü 131 aircraft. One of the fighters turned over on landing as the result of an undercarriage failure. That same day, another five Caproni Ca 135 bombers were also despatched to the front. One of these, flown by *Zászlós* – Cornet – Baracskay, turned back near the Carpathians because of a technical fault. A request for replacements for the damaged aircraft meant that the unit now had ten aircraft. At the beginning of July the fighters made training flights and the bombers some short-range missions of 2-300 kilometres. It soon became evident that the German Ju 88 bomber units could make five to seven sorties daily, but the Hungarian Ca 135 aircraft could make only one or two because of technical problems.

On the 7th July a Ca 135 was lost over Rahó when it caught fire and crashed for unknown reasons, two officers and two NCOs being killed. During July the units of the 2nd Air Brigade were based on the airfields of Sztárij Oszkol and Ilovszkoje. On the 30th July the *I. Vadaszosztály* – 1st Fighter Group – arrived from home, as well as the 2/1. *Vadászszázad* – *2/1.* Fighter Squadron – from Kolozsvár.

The 2nd Hungarian Army arrived at the River Don on the 10th July 1942 and there it had to defend itself. In their sector three Soviet bridge-heads remained: Uriv, Kovotojak and Szcsucsje and attacks soon started against them. The Uriv was the most dangerous spot, three attacks were directed against it; on the 27th July, on the 10th August and on the 9th September, but though the losses were great, the results were not good enough as a bridge-head of 100 square kilometres still remained. Therefore the Hungarian aircraft made more and more attacks, sustaining heavy losses. At dawn on the 28th June an He 46 departed on a mission to observe the Don crossing points and never returned. It simply disappeared, along with the crew of *Zászlós* – Cornet – Frecska and *Főhadnagy* – Lieutenant –

Szily. They did not come back after the war and their fate remains unknown. After this event it was ordered that all reconnaissance aircraft should be escorted by two fighters. Experts stated that there was some magnetic force around Kursk which could have effected the aircraft's compass.

On the 29th June, 1942, the He 111 flown by *Főhadnagy* – Lieutenant – János Hollós and his crew shot down two Soviet fighters in a running battle and returned damaged. On the 2nd July, He 111P, F.7+04, blew up in mid-air after a fighter attack and *Főhadnagy* – Lieutenant – József Benke, the observer; *Hadnagy* – 2nd Lieutenant – László Hechts, the pilot; and their fellow crew members *Tizedes* – Corporal – József Roszkos, photographer; *Szakaszvezető* – Lance Sergeant Jenő Katovics, the gunner and *Tizedes* – Corporal – Ferenc Maróti, the wireless operator, all met their deaths.

On the 4th August there were two air victories above Uriv by the fighters, on the same day eight Soviet fighters attacked an He 111P-6/R2 long distance reconnaissance aircraft above Voronezh. After a fight which lasted 35 minutes the bomber crew shot down three of the attacking aircraft and returned safely to base.

Air activities were intense in August because of the general advance towards Stalingrad. On the 6th, *Hadnagy* – 2nd Lieutenant – Takács of the 1/1. '*Dongó*' squadron was shot down by Soviet fighters. He managed to get the aircraft down, although he was seriously wounded, and landed in the area of the German VI Army. The Héja I was completely destroyed.

On the 6th August, *Főhadnagy* – Lieutenant – István Horthy from the 1/1. Squadron, on patrol with *Őrmester* – Sergeant – Zoltán Nemeslaki on the far side of the Don, met a Soviet unit consisting of 12 Ilyushin Il-2 *Shturmoviks* escorted by 24 LaGG-3 fighters. Making use of their altitude advantage, they bounced the Russians, firing as they flew through the formation and then quickly departed. One LaGG fighter hit by fire from Horthy rolled out from the formation and was claimed as a 'probable'. This was the only victory of the Deputy Regent, István Horthy. After landing, eight bullet holes were discovered in his aircraft. There were also many holes in the aircraft of Sergeant Nemeslaki, who owed his life to his back armour.

Also on this day in the afternoon, *Zászlós* – Cornet – Ferenc Vajda, from the 2/1. Fighter Squadron in the area of Uriv shot down a LaGG fighter and later an Il-2.

On the 9th August an He 46 was attacked by enemy fighters and claimed one shot down. The fighters claimed two shot down.

Two days later the Soviet air defences shot down the *Héja* I fighter of *Őrmester* – Sergeant – Gémes, who took to his parachute and fell into captivity.

Top: *An overall light blue civilian registered Focke-Wulf Fw 58, HA-XBG, used as a personal transport by Vezérőrnagy – Major-General – László Háry in 1943 (Sárhidai)*

Above: *In-flight view of a Fiat CR 42 of the 1/3. 'Kör Ász' – Ace of Hearts – fighter squadron, based at Mátyásföld, near Budapest, in 1942 (Sárhidai)*

Left: *Junkers Ju 87B-2 Stuka dive-bomber used by the MKHL as a trainer at Budaőrs in 1943 (Sárhidai)*

On the 14th August, above Uriv, the Soviet anti-aircraft guns shot down the Ca 135 of *Őrnagy* – Major – István Mocsári, who was the commander of the *4. Bombázóosztaly* – 4th Bomber Group. Mocsári, *Őrmester* – Sergeant – Zoltán Nagy and *Szakaszvezetö* – Lance-Sergeant – Imre Piri were all killed. *Főhadnagy* – Lieutenant – György Orbán, the observer and *Alezredes* – Lieutenant-Colonel – János Németh of the general staff, parachuted, the latter breaking his leg and having to be rescued from no-man's land.

On the 15th August, *Őrmester* – Sergeant – Andor Péterffy, of the *1/1. 'Dongó'* Fighter Squadron, and who had been selected for promotion, is presumed to have been shot down in his *Héja* over Soviet lines by patrolling enemy fighters.

On the 20th August at 05.00 in the morning, an He 46 aircraft of the *3/2. Közelfelderítő század* – 3/2. Reconnaissance Squadron – took off on a mission, followed by two *Héja* I fighters from the *1/1. Vadászszázad* – 1/1. Fighter Squadron. On arriving above the airfield of Ilovszkoje D the escorts attempted to catch up with their charge. It then happened that one of the fighters turned sharply to port at an altitude of 300 metres, got into a tailspin and crashed. Consequently, it was in this aircraft, V-421, that *Főhadnagy* – Lieutenant – István Horthy, the Deputy Regent, met his death. An immediate investigation was ordered, but there was no trace of sabotage or foreign interference. Later many legends arose, but the historical sources and research do not bear these out. The *Héja* aircraft was known to be difficult to handle and during training there were many accidents. When the Deputy Regent died, on his twenty-fifth mission, everybody in the unit and in the country mourned his loss.

As many other machines sustained damage, the German Luftflotte 4 gave three He 111P-6 and four Do 215B-4 aircraft to the reconnaissance group during July, August and September. By the end of the year another He 111 and two Ju 86 courier aircraft had crashed because of technical faults. The German Supreme Command had to release more modern aircraft from their own stock to the Hungarian units at the front, firstly because the Italian types were of indifferent quality and also because of the growing number of tasks confronting the MKHL. Losses amongst German flight crews were by now also heavy.

The land battles went on. Between 25th August and 3rd September 1942, the Axis forces gained a bridge-head at Korotojak, thus the defence line along the Don between the 2nd German and 8th Italian army became 208 km long.

At the beginning of September the *I. Távolfelderítő osztály* – 1st Long-range Reconnaissance Group, reached its 100th mission, but the 1/1. squadron had only three serviceable aircraft. Two of the recently-supplied Do 215 B-4's and an He 111P-6 had been damaged during operations, while the He 111, F.7+10, was only usable for courier services. Another problem arose when autumn arrived early, which hindered the activities

Above: *A courier Messerschmitt Bf 108B 'Taifun', G-3+55, on an unidentified airfield 1942-43. A Bücker Bü 131 can just be seen in the distance (Sárhidai)*

Right: *WM 21 'Sólyom' tactical reconnaissance aircraft, F-2+82, on an Hungarian airfield, sometime in 1943. Unheralded, virtually unknown and not particularly successful, the K-14 engined WM 21 nevertheless served in substantial numbers with the MKHL (Sárhidai)*

Above: *Mass unit take-off by the Junkers Ju 87A dive bombers of an MKHL training unit. Four of these elderly machines were supplied by the Germans. Here, B-6+04 leads the pack off, possibly from Börgönd, in 1943 (Sárhidai)*

of the flying units. On the 24th September the force received some good news, i.e. they had gained their twenty-fifth air victory.

Of these victories, nine belonged to the fighters, one to the bombers, three to the tactical reconnaissance and twelve to the long distance reconnaisance squadrons. On the 15th October the *1. Repülőcsoport* – 1st. Air Division – became the *2. Repülődandár* – 2nd Air Brigade. From the 1st November, *Ezredes* – Colonel – Tibor Fráter, became the new commander. During defensive operations throughout October, when the weather became both rainy and cold, the 3/2. Squadron was recalled, the unit reaching home on the 25th October after losing one aircraft. Their old fabric-covered He 46 machines were not suitable for use in winter conditions, yet in spite of this, the type was successful. During four months of front line service they had gained three air victories.

On the 31st October, 1942, the *4/1. Bombázószázad* – 4/1. Bomber Squadron – finished its front line service. During 1,062 missions they carried out 265 bombing attacks and had dropped 1,700 tons of bombs. Losses were two aircraft shot down, one crashed due to mechanical failure and 11 crew members killed. The *2.* and *3. raj* – 2nd and 3rd flights – returned to Debrecen on the 15th November.

1. raj – 1st flight – stayed as a unit at Poltava, where it received five used He 111 bombers from the Germans and began to train the crews accordingly. Some time before this, the Germans had decided that the 4/1. Squadron would be given Junkers Ju 88 bombers, and therefore unceremoniously took back the He 111's. The air crews subsequently received their training at Istres in France during January-February 1943 under the command of *Főhadnagy* – Lieutenant – Tihamér Ghiczy. Meanwhile, the 2nd Fighter Squadron dwindled to half-size, but one part of it was continuously combat ready. Because of their shrinking combat capability they had only 19 confirmed kills and one probable up to 15th December. Four pilots had died, one had been captured. Total aircraft losses were seven *Héja*. The first MKHL ace was *Hadnagy* – 2nd Lieutenant – Imre Páncél, from the 1/1. Squadron, who by the time of his death on the 11th December, 1943, had shot down five Il-2's. (Three with the *Héja* and two while flying the Messerschmitt Bf 109F).

Air links with the *2. Repülődandár* – 2nd Air Brigade – and the War Office were maintained during 1942 by the six Junkers Ju 52/3m transport aircraft of MALERT.

The air transport squadron was organized by the

main air department of the War Office. The transport pilots frequently flew in very bad weather conditions over the Carpathians on the Budaörs-Berdicsev-Lemberg-Kiev route, up to the airfields at Szstary-Oszkol and Ilovskoje near to the Don. At the end of 1942, the commander of the Air Brigade was based here, along with the *5/2. Vadászszázad* – 5/2. Fighter Squadron, the *1. Szállító repülőszázad* – 1st Air Transport Squadron – and approximately 600 airmen, together with those who were stationed in the area.

After the fighters were re-equipped after the 1st October, *Főhadnagy* – Lieutenant – Bánlaky led the temporary squadron which was formed from the 1/1. and 2/1. squadron crews. This unit gradually received twelve Bf 109F-4 type fighter-bombers to replace the damaged *Héja* I aircraft. The crews were trained on location and the squadron was subordinated to the III. Gruppe of the German JG 52 (III/JG 52) and transferred to Nikolayev. Before the winter set in, the Germans gave four Ju 88D long-range reconnaissance aircraft to their Hungarian allies in the *távolfelderítő* unit. At the same time they declared that modern aircraft would only be available for the Hungarian units in areas where fighting was actually going on, and, furthermore, that they were not transferable. This order was connected with the difficult relationship between the Hungarians and Romanians. When the winter arrived the fighter units were partly non-operational. In the meantime, in Budapest the MKHL high command ordered that the units presently at the front should be withdrawn. Thus, on the 23rd November and on the 15th December the *1/I. Vadászosztály* – 1/I. Fighter Group – command staff and the *2/1. Vadászszázad* – 2/1. Fighter Squadron from Kolozsvár returned home, and the *5/I. Vadászosztály* – 5/I. Fighter Group – command staff and the *5/2. Vadászszázad* – 5/2. Fighter Squadron – went to the front. The 5/2, from Mátyásföld, had only received *Héja* I aircraft in 1942, and training on the type had not then been finished, therefore four pilots from the returning *2/1. Vadászszázad* – 2/1. Fighter Squadron – were left behind as a training cadre. The commander ot the *5/I. Vadászosztály* was *Százados* – Captain – Aladár Heppes who was stationed on the airfield of Ilovskoje. The 2/1. fighter men eventually got their orders to go home on the 2nd January, 1943.

The MKHL units lost 11 of their own aircraft and claimed 63 confirmed and five probable Soviet aircraft kills. As only a few details are known concerning the successes of the air defences, they probably scored more.

To sum up, the air defence artillery of the 2nd Hungarian Army during May-September 1942 shot down 180 Soviet aircraft. In this activity 106 units took part, 7 batteries with heavy guns and a half-unit with 99 automatic weapons. Besides these, from the 25th September to the 19th December, 1942, there are some despatches about 26 more Soviet aircraft. From about the end of this period there is no documentation. During July to September of 1942 the 'Nimrod' self-propelled guns of the Hungarian

Above: *An advanced trainer Arado Ar 96A, G-448, after coming to grief by running into a ditch on an unidentified airfield. The exhaust pipe appears to have been slightly modified (Sárhidai)*

Left: *The white hell of the Eastern Front is graphically displayed as a Junkers Ju 88A-4 of the 4/1. Bombázószázad takes off from an airfield somewhere in Russia in winter 1943-44. Note the propeller spinners which are probably painted in red, white and green, the Hungarian national colours (Sárhidai)*

Left: *One of the old-stagers of the MKHL, Caproni Ca 101/3m, B-113, known as the 'Gavallér' – Cavalier or Gentleman – serving as a paratroop transport with the 1. Szallitószázad at Szombathely in 1942 (Sárhidai)*

army's 51st anti-tank battalion shot down 40 Soviet aircraft with their 40 mm Bofors cannons. There are no documents for other periods. Altogether there were 246 verified Soviet aircraft kills, but if the missing periods are taken into consideration, the final score of the Hungarian air defences would probably be about 260 aircraft.

1943: Disaster on the Don

In January 1943, especially in the first days, the weather was very unfavourable at the front near the Don, consequently the activities in the air were at a low level, also there were no suitable aircraft. The *5/2. Vadászszázad* – 5/2. Fighter Squadron – had just seven usable *Héja*, the *1/1. Vadászszázad* had only 6 Bf 109F's and the four pilots of the *2/1. Vadászszázad* – 2/1. Fighter Squadron – had no aircraft at all. The eight usable Bf 109F aircraft of the 1/1. Fighter Squadron were seconded to the German fighter unit, I/JG 52, for escort and fighter-bomber tasks. The few machines of the long range reconnaissance unit could not take off because of the dreadful weather, therefore they were unable to obtain any preliminary warning of the Soviet attack.

On the 13th January, 1943, the great winter offensive of the Soviet army began – as a result the bulk of the 2nd Hungarian Army was destroyed and the units of the *2. Repülődandár* – 2nd Air Brigade – were surrounded. The encircled units made preparations for a perimeter defence, but as a result of the very cold weather the radial engines of the Héja aircraft would not work. On the 18th January they evacuated the airfield of Ilovskoje, blew up the unserviceable aircraft remaining there and the transport aircraft flew out with maximum loads. On the 19th January, the first day when a breakout was attempted, *Alezredes* – Lieutenant Colonel – Kálmán Csukás and approximately fifty aircrew were killed. Command fell upon *Százados* – Captain – Aladár Heppes, who joined up on the 20th January with the retreating German 26th Westphalian Infantry Division, commanded by Generalmajor Wiese, and on the 21st January, as the rearguard of the German troops moved through Podserednoje they broke out of the encirclement. Aladár Heppes brought with him to Kharkov approximately 1,000 wounded, those vehicles which were in working order and the remaining three air defence automatic guns. After the Don defeat the ground crews amongst these forces returned to Hungary and the air formations were reorganized behind the front.

In January air activities were minimal, only some reconnaissance He 111P-6's flew. At the order of the high command, two aircraft in the front line and a Do 215B-4 which was being repaired in Germany had to return to Budaörs for training. Two of the old He 111P aircraft of the reconnaissance unit had to fulfil courier tasks, while actual reconnaissance missions were to be done with the newer Ju 88D aircraft. All the *Héja* I aircraft of the *5/2. Vadászszázad* – 5/2. Fighter Squadron – had been destroyed, therefore the squadron received the remaining three old Bf 109F fighter-bombers of the 1/1. Fighter Squadron in February. The 1/1. received newer Bf 109F's beginning from the 15th February. These were numbered from V-013 to V-020. From these, V-013 and -014 force-landed in the area of Kharkov and were damaged.

There was a plan for the *1. Szállító repülőszázad* – 1st Air Transport Squadron, then under the

Right: *A Focke-Wulf Fw 58 'Weihe', G-2+51, of the 1. Futárrepepölő raj – 1. Courier Flight – wearing a most unusual brown camouflage scheme for protection as it sits on the 2nd Army landing field at Aleksejevka in 1942 (Sárhidai)*

Below: *An Hungarian PIRT-built Junkers Ju 52/3m transport, possibly S-2+04, sits on an Eastern Front airfield in spring 1944 (Sárhidai)*

Below right: *One of ten Rubik 'Cimbora' – Fellow – gliders used for training possible glider force pilots, seen here near Pápa in winter 1943-1944 (Jávor)*

command of László Ujvári, to use towed gliders as they seemed very useful in the German units, but conditions were never favourable. To replace four machines which had been blown up, Ujvári received four newer Ju 86 aircraft from the stocks left from the disbanded bomber squadrons.

On the 25th March, *Ezredes* – Colonel – Sándor Illy of the general staff took over the command of the *2. Repülődandár*. Even before this, on the 2nd February, reorganization and rearmament of the Brigade had begun. On the 6th April, 1943, they began to pull back the remnants of the 2nd Army, the commander relinquishing his command on the 30th April. Thereafter the *2. Repülődandár* – 2nd Air Brigade – came under German Luftwaffe control. By then, the Hungarian land army had sustained more losses than for the whole year of 1942. Despatches dated 15th March, 1943, referring to the general situation, state that during the fighting in January and February, 75% of the equipment of the *Repülődandár* and the ground formations had been lost. Only 49.3% of the aircraft establishment was still on charge, meaning that 36 aircraft had been lost. Ten *Héja* I aircraft, numbers V-470, -435, -440, -439, -459, -458, -442, -410, -412 and -443 of the *5/2. Vadászszázad* – 5/2. Fighter Squadron – had been blown up, along with the Arado Ar 96, G-525.

Of the Bf 109F's of the *1/1. Vadászszázad* – 1/1. Fighter Squadron – 7 were lost during the retreat, namely V-001, -002, -005, -006, -008, -009 and -010, as well as the two Arado Ar 96's, G-437 and -438. The *4/1 Bombázószázad* – 4/1. Bomber Squadron – lost its Bücker Bü-131 courier aircraft, I-415, while the *1. Szállító repülőszázad* – 1st Air Transport Squadron – lost the following machines:

Junkers Ju 86: B-308 and -346

Focke-Wulf Fw 58: G-251 and G-404, -407

Arado Ar 96: R-108, -109

Fieseler Fi 156: PV+ZT

Bücker Bü-131: I-417

The reconnaissance units lost two Ju 88's, F-902 and -903 and the Fw 58, G-227. It was at this time that Hungary's first ace, *Hadnagy* – 2nd Lieutenant – Imre Páncél died when his Bf 109, V-004, was shot down on 11th January, 1943, by the Soviet air defences.

Characteristically, there was great confusion amongst the maintenance units and repair depots, consequently many aircraft were blown up, thereby unnecessarily increasing the losses.

Casualties amongst the personnel of the Air

Brigade during the Soviet offensives and the break-out from encirclement were also grievous; 36 men being killed and 56 wounded. Another 32 suffered various degrees of frost-bite and 46 simply disappeared.

The German OKW was keen that the Hungarian forces should still take part in the Eastern campaign, but recognized that they would not be in the front line because of the enormous losses of the 2nd Hungarian Army. In the hinterland only the occupying brigades stayed on, at the front line itself only the units of the *2. Repülődandár* remained. The Hungarian political leadership tried to keep only minimal forces on the front line, because the Luftwaffe only released modern aircraft for front line service and thus the flying units could only obtain training experience there. To avoid or minimise bloody losses, they tried to change aircrews of some of the squadrons every six months or so. This was not always possible because the training at home was so limited. In accordance with the Hungarian-German Aircraft Production Agreement, factories were built and production of the Bf 109G, Me 210Ca-1 and Ju 52/3m aircraft and DB 603 engines and instruments by them was accelerating. With these facts in mind, it was necessary to co-operate with the Germans.

The situation at the front on the 1st of March, 1943 was as follows:

The *5/I. Vadászosztály* – 5/I Fighter Group – under Aladár Heppes, by now an *Őrnagy* – Major – had three serviceable Bf 109F's from the *5/2. Vadászszázad* – 5/2. Fighter Squadron, numbers V-003, V-011 and V-012. Additionally, there was the Bü 131, I-421 and the Fw 58 Weihe, G-228, as transports. Nominally under his command was a provisional fighter squadron organized from the *1/1. Vadászszázad* – 1/1. Fighter Squadron – led by *Főhadnagy* – Lieutenant – György Bánlaky, but in reality it was subordinated to the Luftwaffe as part of III/JG 52. There they were involved in operations to defend the route used by the transports heading for Stalingrad.

Earlier, on 15th February, 1943, the *1/1. Vadászszázad* had taken delivery of seven Bf 109F-4 fighters; numbers V-007, -013, -016, -017, -018, -019 and -020. Later the Germans raised this to twelve aircraft. Also handed over was an Ar 96, G-431, and a Bü 131, I-478, as courier aircraft.

At this time the *4/1. Bombázószázad* – 4/1. Bomber Squadron – had only a single Fw 58 Weihe, G-230. They also received five German He 111 bombers in the interim, but these were handed back without compensation, and the crews were removed from the front to commence re-training on the Ju 88 in France.

The *1. Szállító repülőszázad* – 1st Air Transport Squadron – had only three Fi 156 Storchs; R-116, -102, -104 and the damaged R-107. Also on strength was Bü 131, I-444 and three Fw 58's; G-228, -247 and -230, which was damaged. The *I. Távolfelderítő osztály* – 1st Long-range Reconnaissance Group – was active under the command of *Őrnagy* – Major – Elemér Kovács. Under him was the *1. Távolfelderítő század* – 1st Long-range Reconnaissance Squadron commanded by *Főhadnagy* – Lieutenant – Adorján Mersich attached to the Luftwaffe Kommando Don. The squadron had two He 111's; F-703, and -706; and nine Ju 88D's; F-901, -904, -905, -906, -907, -908, -909, -910 and -911. These arrived after the 20th December 1942, as did one Bü 131: I-423. During the earlier battles, one Ju 88, F-902, was damaged on landing, while F-903, -912 and -913 were destroyed

Right: *Dramatic in-flight shot of a group of Hungarian Junkers Ju 87D dive bombers of the 2/1. Zubo század seen from the rear-gunner's seat of an accompanying aircraft. Somewhere over the Eastern Front in summer 1944 (Sárhidai)*

on the ground. Amongst the older aircraft, an He 111, F-710, remained at Budaörs, while He 111, F-705, and the Dornier Do 215, F-707, were taken back to Budapest for training purposes. The Headquarters staff of the occupation forces used Bü 131, I-504; Ar 96, G-429; two Fi 156 Storches, R-112 and -114, and a captured Soviet Polikarpov Po-2, I-144.

Conversion of the *5/2. Vadászszázad* – 5/2. Fighter Squadron, *'Puma'* – onto the more modern Bf 109G-2 fighters began in March 1943 in Kharkov and by April the 5/2. and 1/1. Squadrons of the *5/I. Vadászosztály* – 5/I. Fighter Group – were operational. *Főhadnagy* – Lieutenant – Gyula Horváth became commander of the 5/2. and *Főhadnagy* György Ujszászy that of the 1/1. *Főhadnagy* – Lieutenant – György Bánlaky served as the unit adjutant until April, 1943, when he returned to Hungary.

The squadrons were used mainly for fighter-bomber tasks loaded with 50kg and 250kg bombs. On occasions they were used for escorting the German Ju 88 and He 111 bomber units, when they employed a mix of Bf 109F and G types.

In November 1943 the command staff of *5/I. Vadászosztály* – 5/I. Fighter Group and the *1/1. Vadászszázad* – 1/1. Fighter Squadron, without aircraft, were ordered back to Hungary. The *5/2. Vadászszázad* – 5/2. Fighter Squadron under the command of *Főhadnagy* – Lieutenant – Gyula Horváth was given the left-over aircraft and was thus brought up to full combat strength.

In September *Őrnagy* – Major – Aladár Heppes was ordered back to Budapest and the name of the 5/2. Fighter Squadron was changed to the *Önálló Vadászszázad* – Independent Fighter Squadron. On the 5th October the 2nd Air Brigade became the *102. Repülődandár* – 102nd Air Brigade, when the 5/2. was renamed the *102/1. Önálló Vadászszázad* – 102/1. Independent Fighter Squadron and given a new commander in the person of *Százados* – Captain – József Kovács.

The *5/I. Vadászosztály* – 5/I. Fighter Group carried out 1,560 missions on the Eastern front between April and November, 1943. Meanwhile, the returned *Puma* squadron lost nine aircraft and six pilots and claimed more than 60 air victories. Later records state that until December 1943 the two squadrons, combined, lost nine pilots of whom six were missing. Six aircraft disappeared, three were damaged and another twenty were wrecked. 132 victories were claimed.

After retraining, the *1/1. Távolfelderítő század* – 1/1. Reconnaissance Squadron – carried out more than 1,000 missions in increasingly frequent battles in the central sector of the front line between June 1943 to January 1944. In this period five of their Ju 88D aircraft were shot down, leaving them with only one undamaged combat-worthy machine, all the others being damaged. During this period they claimed 23 air victories. In November 1943, as the Air Brigade command staff were disbanded, the reconnaissance squadron was re-christened the 1st Long Range Reconnaissance Squadron. The *I. Távolfelderítő osztály* – 1st. Reconnaissance Group commander, *Őrnagy* – Major Elemér Kovács had already returned home in June 1943. His successor was *Őrnagy* Gyula Timár. By now the unit was acting completely under German control.

The situation of the bomber units was very difficult because of the lack of suitable aircraft. In February 1943, the War Office had cancelled the

***Left:** Aircrew from an unidentified bomber squadron equipped with the Junkers Ju 86K take a break prior to a mission from Veszprém in 1943. Note the well-worn camouflage paint and the unfinished fuselage marking which consists only of the white cross (Jávor)*

Focke-Wulf Fw 190F-8, W.521, of the *102/2. Vadászbombázó század* – 102/2. Fighter-Bomber Squadron of the *102. Csaterepülő osztály* – 102nd Air Battle Group – was flown by section leader Sándor Horváth from the airfield at Várpalota in winter 1944-1945. The yellow 'V' tactical marking on the wing was introduced for all Axis aircraft operating in Hungary on 28th September 1944. It was later superseded by an order from Fliegerkorps IV on 7th March 1945 which required 50 cm wide yellow bands on the nose and fuselage, and a similar coloured rudder

Messerschmitt Bf 109G-10, Blue 12, was a German-supplied machine of the *101. Vadászezred, I. Osztály* – 101st Fighter Regiment, 1st Group – which was lost in a collision with another aircraft flown by *Főhadnagy Báró* – Lieutenant, the Baron – Pál Luzsénszky on 16th March 1945, owing to the pilot being wounded. Luzsénszky seems to have been a singularly unfortunate pilot, being forced to make a crash landing on the 22nd February and being shot down over Vienna on 14th April

Another German supplied machine, **Messerschmitt Bf 109G-14**, Werk No. 611022, Red 5, of the 102/1. *'Ricsi'* squadron, whose emblem can be seen on the engine cowling. This was the aircraft of *Százados* – Captain – László Pottyondy, the squadron commander. The aircraft survived until the 2nd May 1945

Messerschmitt Bf 109G-14/AS, Yellow 15, of the 101/III. i.e. *101. Vadászezred, III. Osztály* – 101st Fighter Regiment, 3rd Group, seen at Raffelding in April 1945. Note the yellow rudder and nose band, markings adopted by Luftflotte Reich in early 1945

Messerschmitt Bf 109G-14, W-058, an Hungarian-built machine which was allocated a German Werknummer, 282205, and seen here at Veszprém in September 1944. At the time it belonged to the *101. Vadászezred, 5 Század* - 101/5. i.e. 101st Fighter Regiment, 5th Squadron, and was flown by *Hadnagy* – Captain – Antal Szebeni

Messerschmitt Bf 109G-6, V.847, white 6, seen at the Győr factory airfield in September 1944 when it belonged to the *101/I. Vadászezred, I Osztály, 2. Század* – i.e. 101st Fighter Regiment, 1st Group, 2nd Squadron. At that time it was flown by *Hadnagy* – Second Lieutenant – Tibor Tobak

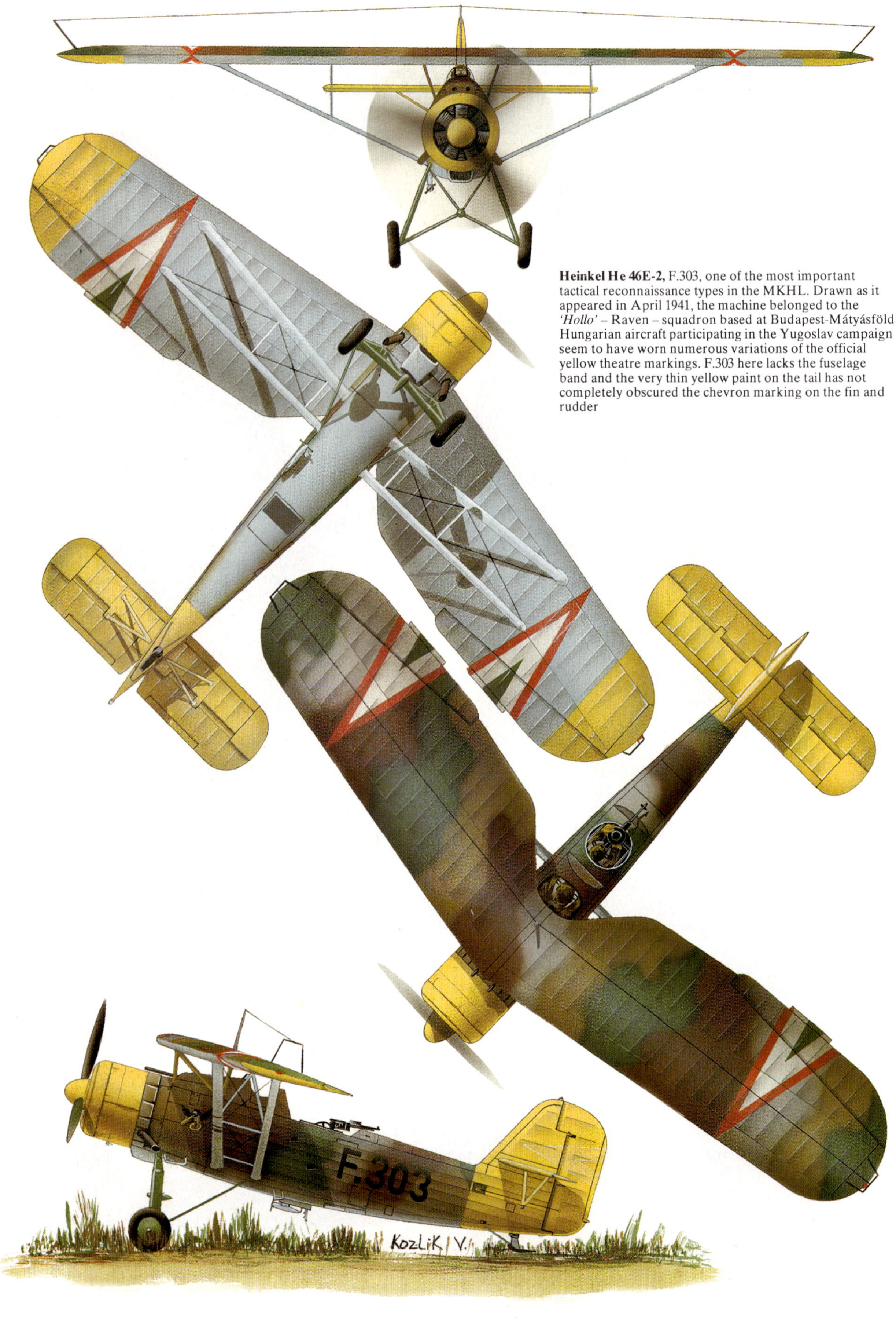

Heinkel He 46E-2, F.303, one of the most important tactical reconnaissance types in the MKHL. Drawn as it appeared in April 1941, the machine belonged to the *'Hollo'* – Raven – squadron based at Budapest-Mátyásföld. Hungarian aircraft participating in the Yugoslav campaign seem to have worn numerous variations of the official yellow theatre markings. F.303 here lacks the fuselage band and the very thin yellow paint on the tail has not completely obscured the chevron marking on the fin and rudder

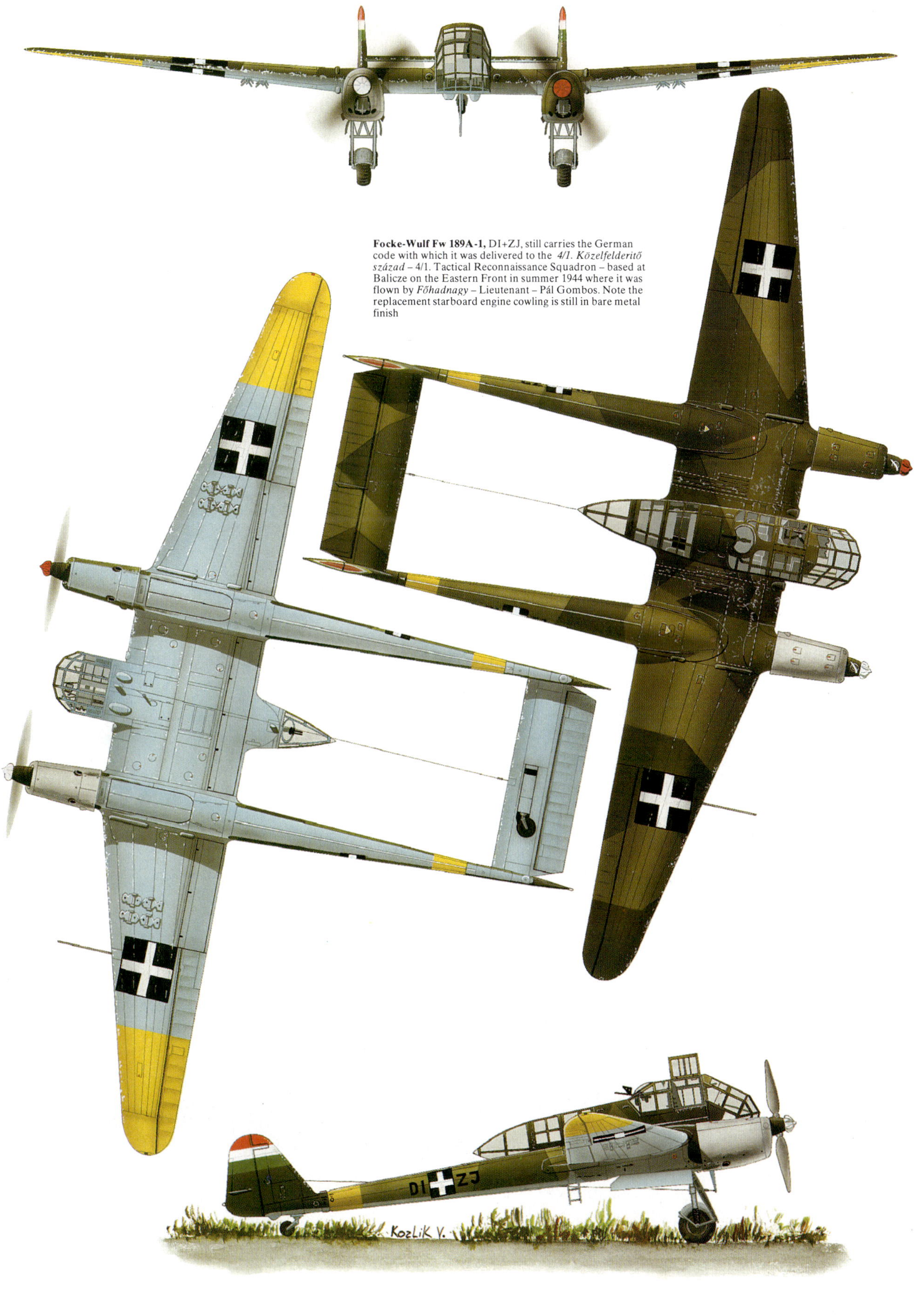

Focke-Wulf Fw 189A-1, DI+ZJ, still carries the German code with which it was delivered to the *4/1. Közelfelderitő század* – 4/1. Tactical Reconnaissance Squadron – based at Balicze on the Eastern Front in summer 1944 where it was flown by *Főhadnagy* – Lieutenant – Pál Gombos. Note the replacement starboard engine cowling is still in bare metal finish

Junkers Ju 88A-4, B.142, of an unidentified unit, carrying a well-worn winter camouflage finish, somewhere on the Eastern Front in winter 1943

Dornier Do 215B-4, F.708, of the *2. Repülődandár, 1. Önálló távolfelderítő osztály, 1/1. század* i.e. 2nd Air Brigade, 1st Independent Tactical Reconnaissance Group. 1st Squadron. The Do 215 was used by both the Hungarians and the Luftwaffe for reconnaissance tasks. This particular machine is shown as it appeared in August 1942 on the airfield at Amaszovka on the Eastern Front

Junkers Ju 87D-3, ZB+00, of the *102/2. Zuhanóbombázó század* – 102/2. Dive Bomber Squadron – based at Kiev in April 1943.

Fiat CR 42, V.239, of the *2/4. Vadászszázad* – 2/4. Fighter Squadron – the 'Arrows', probably at Kolozsvár in early 1940. Note how the fuselage codes have been hastily overpainted

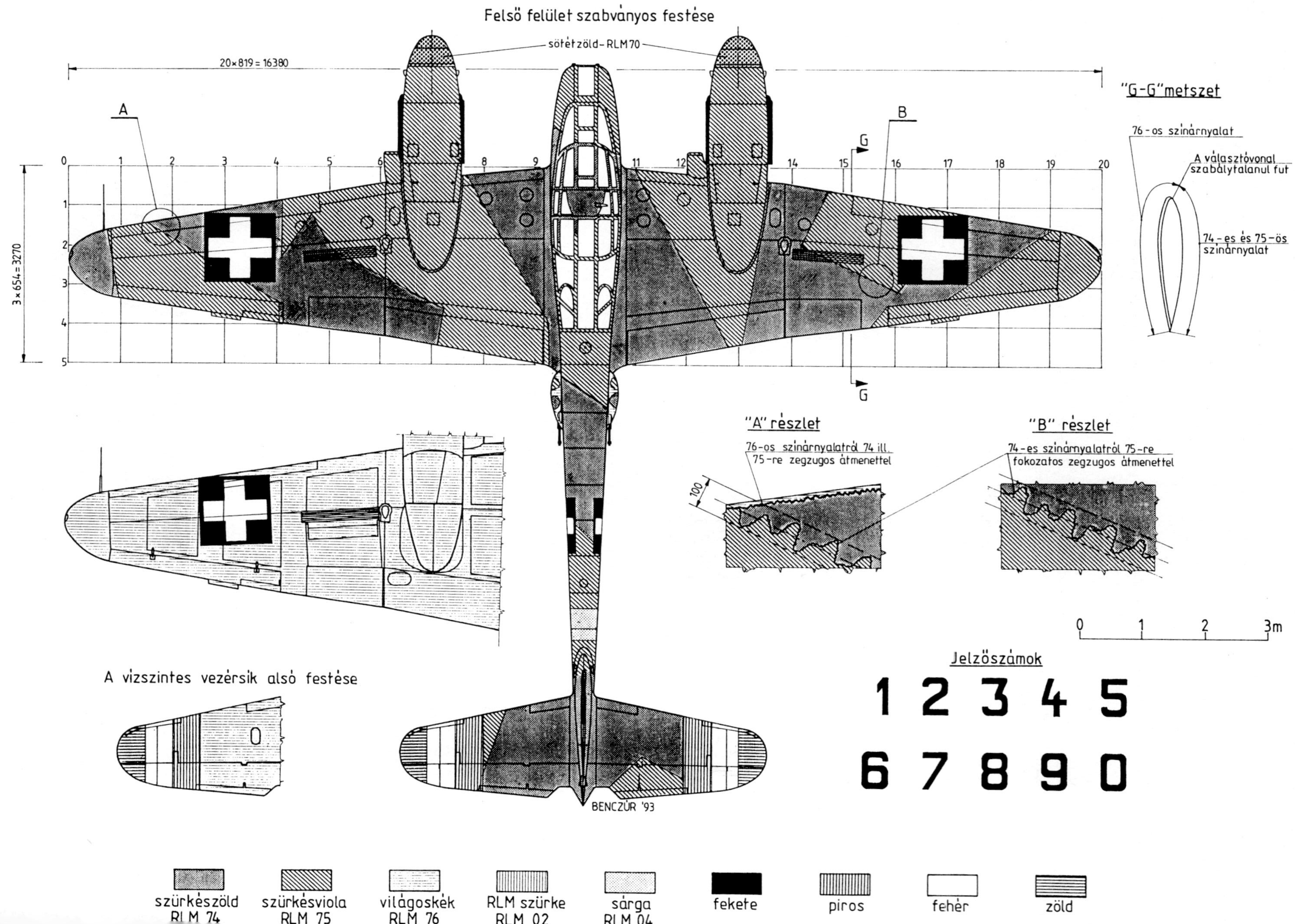

Felső felület szabványos festése
sötétzöld-RLM70
20×819=16380
3×654=3270
A
B
G
"G-G" metszet
76-os színárnyalat
A választóvonal szabálytalanul fut
74-es és 75-ös színárnyalat
"A" részlet
76-os színárnyalatról 74 ill. 75-re zegzugos átmenettel
100
"B" részlet
74-es színárnyalatról 75-re fokozatos zegzugos átmenettel
0 1 2 3m
Jelzőszámok
1 2 3 4 5
6 7 8 9 0
A vízszintes vezérsík alsó festése
BENCZÚR '93
szürkészöld RLM 74
szürkésviola RLM 75
világoskék RLM 76
RLM szürke RLM 02
sárga RLM 04
fekete
piros
fehér
zöld

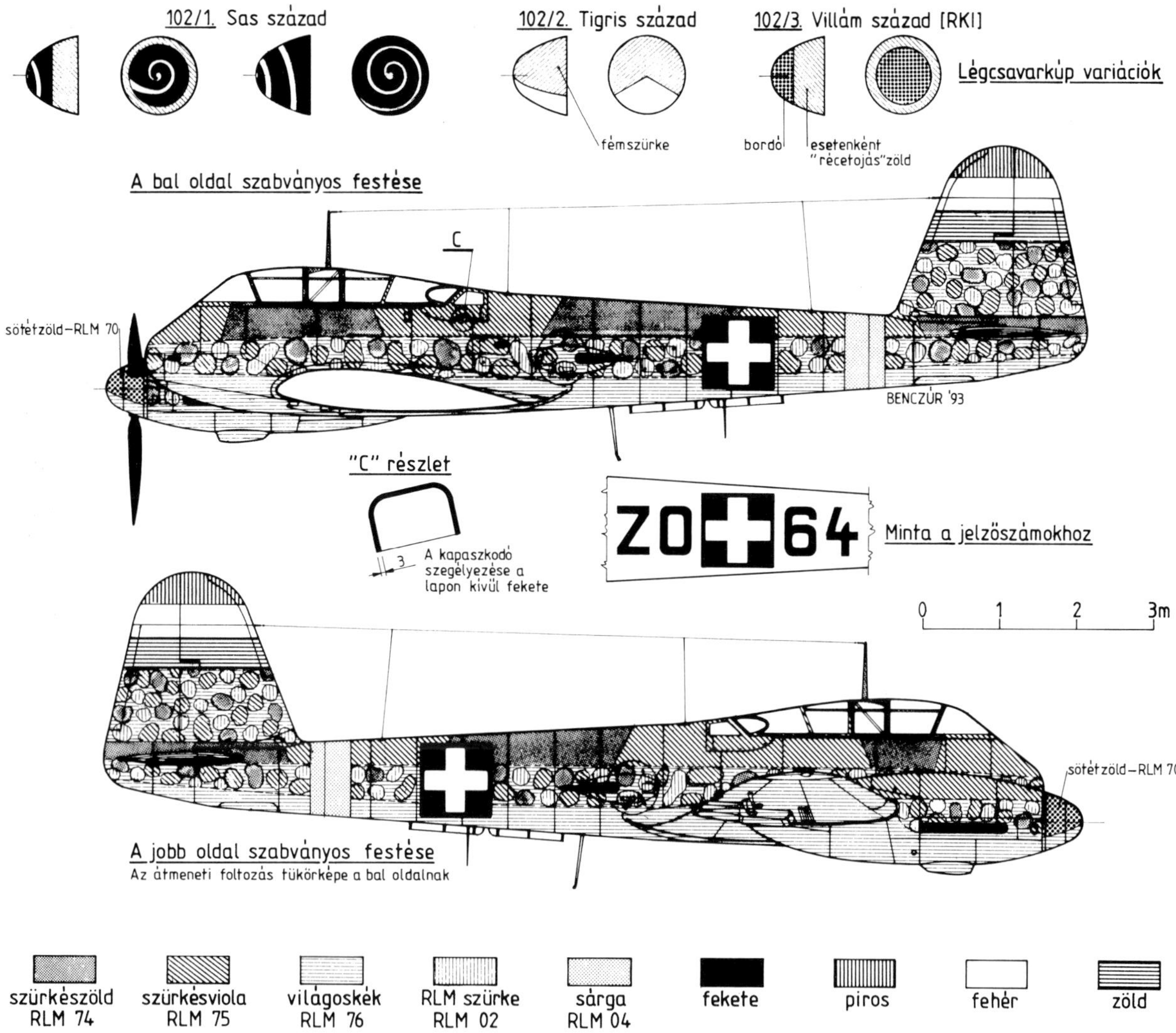

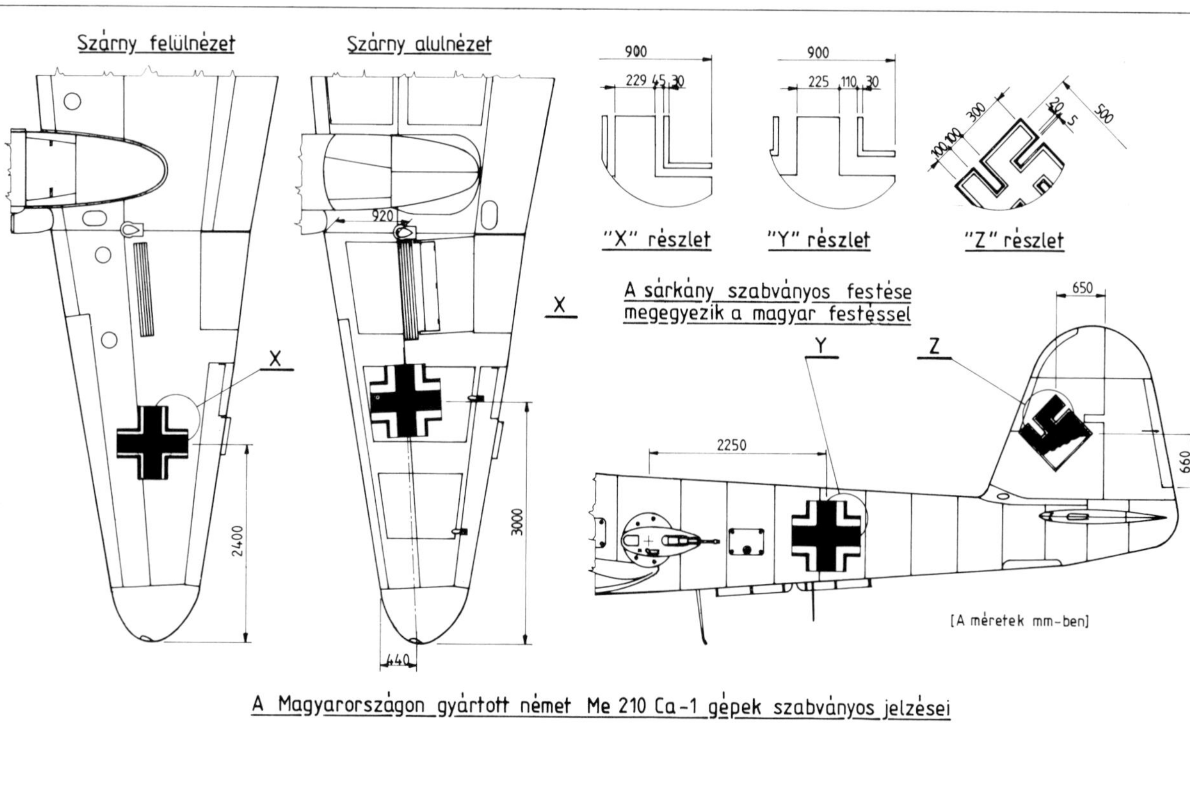

A Magyarországon gyártott német Me 210 Ca-1 gépek szabványos jelzései

1 2 3 4 5 6 7
LFA
8 9 10 11 12 13 14
15 16 17 18 19 20 21
22 23 24 25 26 27 28

Right and Left: *It is very unusual to find views of both sides of the same aircraft. Fiat G.12, S-1+52, was the second of eighteen delivered to the MKHL where, amongst other tasks, they were used to carry paratroops with the 102/2. Ejtőernyős szállitószázad. The aircraft had apparently seen earlier service with both the Regia Aeronautica and the Luftwaffe. On the port side view, taken at Pápa in 1944, the lightly overpainted Savoia cross can be still be seen on the rudder. On the starboard side view, also taken in 1944, there are what appear to be two yellow fuselage bands, typical of those used by Luftwaffe blind-flying trainers. Note also the 'W-51' code, the origin of which is unknown (Sárhidai)*

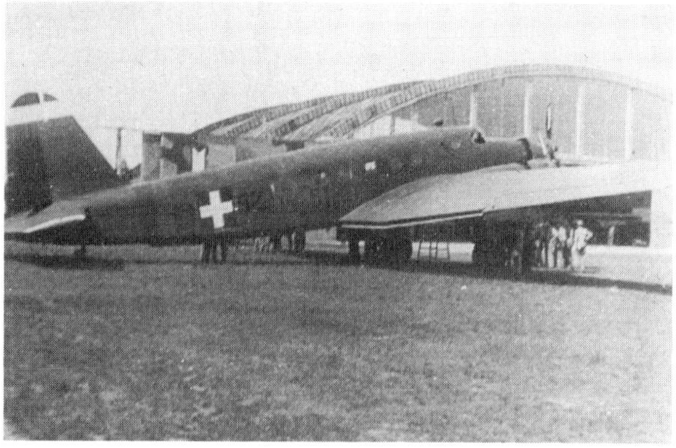

second delivery of 36 of the mediocre Italian Caproni Ca 135 bis bombers, as superfluous to requirements, but the aircraft still arrived during 1943 from Regia Aeronautica stocks. At this time the type was completely obsolescent and could only be used for training at home. During spring of 1943 the crews of the 4/1. and the 3/1. Bomber Squadrons were sent from Hungary for retraining at the Istres training school near Marseille in France. The latter unit, renamed *102/1. Bombázószázad* – 102/1. Bomber Squadron, under the command of *Százados* – Captain – László Herder, and equipped with 12 Ju 88A-4 aircraft was deployed in the central area of the front under the command of a German Ju 88 unit. From April 1943 they flew more than 1,000 missions, thereafter they were taken out of the front line and returned to Hungary without their aircraft. At the end of the year the remaining machines were given to the *1. Távolfelderítő század* – 1st Long Range Reconnaissance Squadron.

Meanwhile, the newly established *102. Bombázóosztály* – 102nd Bomber Group's second squadron was named *102/2. Zuhanóbombázó század* (usually abbreviated to *zubó*) *'Kókuszdió'* – 102/2. Dive-bomber 'Coconut' Squadron – equipped with twelve Junkers Ju 87D-3 dive-bombers under the command of *Százados* – Captain – János Kórosy. This was sent out to the front in July and took part in the battles from August until October in the region of Kharkov. In just short of three months as a squadron of the German Stukageschwader, it carried out 1,500 sorties, losing eight aircraft to Russian ground fire. These either crashed in their own lines or had to land, many of the crews being wounded. Because there were only four operational aircraft and crews left, the squadron was ordered back to Kolozsvár to train crews and fill the gaps. In the official register of the MHKL it was designated *2/2. Zuhanóbombázó század* – 2/2. Dive-bomber Squadron.

In March 1943 the *3/1. Közelfelderitő század 'Táltos'* – 3/1. Tactical Reconnaissance Squadron, under the command of *Százados* – Captain – Imre Telbisz went to the Soviet front, equipped with old WM-21 *Sólyom* aircraft. Once there, these were exchanged for twelve modern German Focke-Wulf Fw 189's. After a brief conversion course, the unit was deployed in the offensive against Kharkov, where these twin-engined machines with their slender twin tail-booms, which offered poor targets for the air defence guns, were quickly nicknamed *'Bel Ami'*. In the air, three machine guns protected them from attacks by the Soviet fighters. Their first mission was on the 22nd May 1943, and from that time on they carried out constant reconnaissance and harassing bomber activities until March 1944. The unit lost only two aircraft and remained effective until the very end.

In the winter of 1943-44, the aircraft losses at the front were made up by the Luftwaffe, while at the same time the air force back home in Hungary tried to create more units with more modern equipment from the slowly developing Hungarian production programmes. At the end of 1943 two great centres for these were being built up. In Budapest, or more exactly near Budapest, in Szigetszentmiklós, was the so-called Danubian Aircraft Factory; the Weiss Manfred Aircraft Engine factory at Csepel and MÁVAG at Győr. Thus the *5/I. Vadászosztály* – 5/I. Fighter Group – at Mátyásföld reorganized the 5/1. Fighter Squadron into the *5/1. Kísérleti Éjszakai Vadászszázad* – 5/1. Experimental Nightfighter Squadron under the command of *Százados* – Captain – Ádám Krudy, and moved them to the newly finished airfield at Ferihegy.

Meanwhile, the Kállay Government took cautious steps to reduce Hungarian war obligations and tried to preserve a larger military force intact for the last and this time, predictable, period, namely the expected Soviet counter-offensive. Consequently neither army nor air force units were directed to the front.

For night training in 1943 there were only three radio-equipped CR 42's, three Bf 110F heavy fighters, and an assortment of Fw 58, Ju 86, Ar 96 and Ro 41 types. From the beginning of 1944, to these were added some nine to twelve Me 210E night fighters and three more Bf 110G-2's. The *Repülő Kísérleti Intézet - RKI* – Air Experimental Institute – gave orders to *Őrnagy* – Major – Lóránt Dóczy at the end of the summer that he should form a unit with at least 12-15 aircraft using the Me-210Ca-1 'destroyers' which were then at Ferihegy for testing. Under his command, the test pilots from the Institute would fly these aircraft and thus could be used for operations in case there was an air attack during the day. Organisation of this unit into a Staff, three combat flights and a reserve flight began, with a total establishment of fifteen aircraft, but it never reached operational status.

Above: *A Focke-Wulf Fw 56 'Stösser', G-1+4?, in service as a hack with the 102. Vadászostály – 102nd Fighter Group – at Szamosfalva in May 1944. Note the dark finish and the unit's 'Ricsi' emblem (Ketley)*

Above right: *A CR 42 fighter in use as a trainer with the 1/1. Vadászszázad – 1/1. Fighter Squadron at Szolnok in 1942 (Sárhidai)*

Right: *A poor quality, but rare, picture of one of the Messerschmitt Bf 110F's leased from the Germans for nightfighter training at Veszprém in late 1943. Note the black-outlined yellow number and spinners (Jávor)*

HUNGARY BECOMES A BATTLEFIELD

The MKHL on the Eastern Front in 1944

Above: *An overall bright blue Messerschmitt Bf 108 'Taifun', G-3+57, flies over the Margit Bridge in Budapest in summer 1942, during the relatively peaceful days before the Soviet onslaught (Sárhidai)*

From January until March 1944, the only Hungarian aircraft on the Soviet front were the five or six Ju-88D's from the *'Hétmérföldes csizma'* – 'Seven-league Boots' – long range reconnaissance squadron; the 12 Bf 109G-6's of the *102/1. Őnálló Vadászszázad* – 102/1. Independent Fighter Squadron and the 12 Fw 189A's of the *3/1. Közelfelderitő század 'Táltos'* – 3/1. Tactical Reconnaissance Squadron, manned by aircrews in rotation. The commander of the fighter squadron was *Százados* – Captain – József Kovács. *Százados* József Frauenhoffer commanded the reconnaissance squadrons. After January the *102/1. Őnálló Vadászszázad* was enlarged with the addition of the fighter squadron from Kolozsvár, thus the *102. Vadászosztály* – 102nd Fighter Group – was created under the command of newly-promoted *Őrnagy* – Major – Kovács. This now consisted of the *102/1. Vadászszázad* – 102/1 Fighter Squadron – under *Százados* – Captain – László Máthé and the *102/2. 'Ricsi'* Fighter Squadron commanded by *Százados* László Pottyondy. Each squadron was equipped with 12 Bf 109G-6 fighters.

According to a document from the OKW, from the summer of 1942 until 1st March 1944, the German Luftflotte 4 handed over to the Hungarian units within their operational area the following combat equipment: Eight Do-215B-4 long distance reconnaissance aircraft, numbered F-750 to 757; 51 Ju 88A-4, Ju 88C-6 and Ju 88D bomber and reconnaissance aircraft; 66 Bf 109 F, G2 and G6 fighters and fighter-bombers; 13 Ju 87D-3 and D-5 dive-bombers; 16 Fw 189A tactical reconnaissance aircraft and ten He 111P long distance reconnaissance aircraft. From these last, five were returned. Luftflotte 4 also gave an unspecified number of Fi 156 and Bü 131 courier and liaison machines for tasks behind the front.

From the 66 Bf 109's there were 12 F-4B's, 40-41 G-2's, and 13-14 G-6 variants.

In May 1944 the *102. Vadászosztály* – 102nd Fighter Group – was under the command the German *Fliegerführer Ungarn* as far as military operations were concerned. This time only two military operational units were left. On all front lines the units were subordinated to the *102. Vadászosztály* under the command of *Vezérőrnagy* – Major-General – Aladár Szirmay.

All Hungarian aircraft units were re-numbered from 101, and the Hungarian fighter squadrons came under the command of the Air Defence Forces HQ headed by *Altábornagy* – Lieutenant-General – Emil Juszthy.

The *2. Repülődandár* – 2nd Air Brigade – came directly under the supervision of Luftflotte 4, but the *2. Szállító repülőszázad* – 2nd Air Transport Squadron – equipped with Ju 52/3m and Ju 86 aircraft, remained strictly under the direct control of the Air Department of the War Office.

This was the situation prevailing at the time of the German occupation of Hungary on the 19th March 1944. At that time all the units at the front line were assigned to the VIII Fliegerkorps, while the remaining units based in Hungary itself came under the command and control of Luftflotte 4. At the end of March 1944, the remaining personnel of the *1/102.* and *3/102. Közelfelderítő század* and the *1/102. Távolfelderítő század* – reconnaissance squadrons were pulled back from the front line, without aircraft, and returned home. At this time a very complicated situation arose. The newly formed Sztójay Government ordered increased involvment in the war, thus in April it was ordered that the 1st Hungarian Army should be mobilized and sent to the Eastern Front. This was followed by the 2nd Army in the summer, which took up defensive positions in the mountains and passes to the east of the Carpathians, but with insufficient resources along the long frontier of northern Transylvania facing Romania. These two armies, which consisted of approximately 600,000 soldiers had no air forces of their own. The *102. Repülődandár* – 102nd Air Brigade – had only two fighter, one tactical reconnaissance and one dive bombing squadron, the long distance reconnaissance and bombing squadrons without aircraft being called back from the front.

Then, following the re-training of the crews of these units, the 1st and 2nd squadrons of the *102. Gyorsbombázó osztály* – 102nd Fast Bomber Group – were formed in May and June 1944. The *102/1. 'Sas'* – Eagle – Squadron was formed in Klemansovo, the *102/2. 'Tigris'* – Tiger – Squadron was formed in Hajduböszörmény. The former was under the command of *Százados* – Captain – Kálmán Háry, the latter under *Százados* Pál Bodó. *Őrnagy* – Major – Albert Martini became Group Commander. From the end of August 1944 these two squadrons, each with 12 Me 210Ca-1 heavy fighters were available in support of the land battles.

Air war over Hungary

Exploratory RAF flights in the airspace over Hungary began in the summer of 1942. On the 4th, then on the 9th September, Soviet long distance bomber units unexpectedly attacked the Hungarian capital. The greatest effect was upon morale. It was made clear to the population and the military leadership, that although the ground battles were faraway on the Don and at Stalingrad, the air space of the country was open and the cities could be bombed.

Between 14th-26th January 1943, the leaders of the Allied powers made a decision at the Casablanca conference that, starting from North Africa, then from Southern Italy, a second "air front" would be created, To do so they would form a

Left: *Seen from the cockpit of an accompanying aircraft, Hadnagy – Second Lieutenant – Mihály Ónodi pilots his Junkers Ju 88A-4, B-1+49, of the 102/1. Bombázószázad – 102/1. bomber Squadron – somewhere over the Eastern Front in 1944 (Sárhidai)*

new air force and its duty would be the bombing of South and Eastern Europe. Within this area Austria, Hungary, Czechoslovakia and the Balkans would be the targets. Results were anticipated by March 1944. Bombing of Hungary's neighbours – Romania and Austria – began in August 1943, and although the attacking units did not yet drop bombs on Hungary, they flew through Hungarian airspace.

On the 24th August, in Quebec, the Allies decided that they would expand their plans for a Combined Bomber Offensive against central Europe. Firstly from North Africa and by the end of 1943, in southern Italy, a new air force, the 15th USAAF, began operations. On the 14th November, 1943, the 15th USAAF command ordered that Budapest, Csepel and Győr should be bombed. By 31st March 1944, this organisation had 21 bomber and 7 fighter groups which meant 1,500 heavy bombers and 800 fighters. In the meantime, on the 19th March 1944, following tentative attempts by the Horthy Government to initiate peace talks with the Allies, German forces occupied Hungary and a new government, completely subservient to the Germans, was formed.

From watching the situation on the Italian front it was expected that the bombing would begin in earnest by March 1944. That these assumptions were correct was verified by over-flights in March and the beginning of the bombing on the 3rd April. Steps had been taken earlier to increase the Hungarian fighter forces and air defences, viz: the transformation in July, 1943, of the 5/1. Fighter Squadron at Ferihegy into an experimental night fighter squadron and the addition of the *2/1. Vadászszázad* – 2/1. Fighter Squadron – with *Héja* II aircraft to the air defences of Budapest. This squadron, with 18 aircraft, moved in October 1943 from Mátyásföld to Ferihegy under the command of *Százados* – Captain – Miklós Scholtz.

The *1/1. Vadászszázad* – 1/1. Fighter Squadron – with 12 *Héja* II fighters, which had formerly been training in Szolnok, was also to be deployed in defence of the capital when necessity arose. At the beginning of January 1944, the *2/1. Vadászszázad* – 2/1. Fighter Squadron – took delivery of 16 Hungarian produced Bf 109G-4 and G-6 fighters. In March, the *1/1. Vadászszázad* received six Bf 109G-6's.

Also in March 1944, the first steps were taken to organise a squadron in Ferihegy, equipped with 18 Bf 109G aircraft, because one fighter squadron was insufficient for the air defence of Budapest. This was the *5/3. Vadászszázad* – 5/3. Fighter Squadron, which later became the 3rd squadron of the 101st fighter group in Veszprém after the 1st May 1944.

Available fighter forces, early April 1944

Unit/Base/Type

2/1. Vadászszázad – 2/1 Fighter Squadron, Ferihegy
12 Bf 109G4/G6

1/1. Vadászszázad – 1/1 Fighter Squadron, Szolnok
4 *Héja*-II + 4-6 Bf 109G-6

1/2. Vadászszázad – 1/2 Fighter Squadron, Szolnok
4 *Héja*-II

5/1. Éjszakai vadászszázad – 5/1 Nightfighter Squadron, Ferihegy 6-10 Me 210Ca-E

RKI Romboló század – Air Experimental Institute Destroyer Squadron, Ferihegy 12 Me 210Ca-1

5/3. Vadászszázad – 5/3 Fighter Squadron, Tököl
2 Bf 109G-6

RKI Vadászszázad – Air Experimental Institute Fighter Squadron, Ferihegy 4-6 Bf 109G-6

RÁB Készültség – Aircraft Acceptance Unit Alert Flight, Tököl 2 Me 210Ca-1

RÁB Készültség – Aircraft Acceptance Unit Alert Flight, Győr 2 Bf 109G-6

These forces were totally inadequate; the 5/1. Nightfighter Squadron was not ready for combat and its use by day was senseless. *The Repülő Kísérleti Intézet (RKI)* – Air Experimental Institute – had two squadrons, but both had pilots who had no military experience; they were test pilots. Equally, the aircraft of the *Repülő Átvételi Bizottság (RÁB)* – Aircraft Acceptance Units – alert flights were flown by test pilots from the factories. This was shown in the very bloody losses during April, when the out-of-date *Héja* II aircraft were clearly only good enough for training and not for operations.

Thus, in practice, one combat-worthy, and two under-trained fighter squadrons of the Hungarian air forces began the fight against the attacking 15th USAAF forces. On the 1st May 1944, these three squadrons at last received 12 Bf 109G-6 fighters each and were given a place in the *101.*

Vadászosztály – 101st Fighter Group, the *'Pumas'*, under the command of *Őrnagy* – Major Aladár Heppes in Veszprém. From here they took off and contested the American forces during the summer.

The first clash between Hungarian and American aircraft happened on the 17th March. On the morning of 3rd April, the first really big American attack reached Budapest and its surrounding area. 1,036 people died.

That same night the bombers from the RAF's 205 Group repeated the attack. The Allied air forces thus began the air campaign against Hungary. In the day time the 15th USAAF, at night 205 Group of the RAF, attacked targets throughout the country. The capital, Budapest, suffered 26 heavy air raids. Over half of these were carried out at night by the RAF. Almost the same fate happened to most of the big cities, the industrial basis of the country. To summarise, the aims of the Allied air forces were the following:

To destroy and paralyze the rail traffic of the country, which was the most important link between Germany and the southern part of the Eastern Front in the Soviet Union as the modern railway system was well developed in Hungary and transported much of the petroleum from Romania to Germany. 205 Group, RAF, also hindered transportation along the Danube with mine-laying raids.

From the beginning of 1944, the destruction of

Above: *Taken at the war's end, at Neubiberg, this picture shows some of the last Messerschmitt Bf 109G fighters of the MKHL. Nearest is Werk No. 613165, 'Black 16', a Wiener-Neustadt built G-10 which once belonged to the 101/II. Vadászosztály of the 'Pumas' Fighter Regiment. Note the late-war style black outline crosses on the wing, and the yellow nose band and rudder. At least two more Hungarian 109's can be seen (AMC via Ketley)*

the Hungarian aircraft production capability, which was becoming increasingly important; to eliminate the ability to manufacture tanks and destroy the petroleum industry.

To break the morale of the population.

The Hungarians opposed these attacks with all the forces at their disposal, and during August 1944 made strenuous efforts to disrupt the desperate missions by the RAF crews overflying Hungary on supply missions to help the uprising in Warsaw. Although the American, British and Soviet air forces, from the beginning of August 1944, caused great damage, the conclusion must be that their air forces, like the MKHL, were insufficiently powerful to achieve their aims.

The last days

After the 1st September, 1944 the Hungarians succeeded in enlarging the three squadron strong *101. Vadászosztály* – 101st Fighter Group – into a six-squadron ezred – regiment. The 101/6. began activities in Kenyeri as a training squadron at this time, while the 101/4. and 101/5. squadrons, also in Kenyeri, reached operational readiness in November. These formed the *101/II. Vadászosztály* – 101/II. Fighter Group, while the *101/I. Vadászosztály* which had the 1st, 2nd and 3rd squadrons stayed in Veszprém. Officially, the two-group regiment began operations on the 19th of September, 1944, in accordance with the orders of *Vezérőrnagy* – Major-General – Sándor Illy, who was the commander of the *1. Repülő Hadosztály* – 1st Air Division. From the 12th October, 1944, the regiment, plus the 5/1. Nightfighter Squadron, was put under the control of the Luftwaffe's 8 Jagddivision. By this time a single command already controlled operations both over Hungary and at the front lines.

During the autumn months, missions against the American formations greatly decreased, as every useable unit and drop of fuel was needed for use against the encroaching Soviet air forces and armies. By autumn the second squadron of the *102. Vadászosztály* was already fighting over Hungarian territory all the time, therefore at the beginning of 1945, during the siege of Budapest, it was decided that all the fighter units should be placed under the same command.

The Hungarian Government and the high-level military command structure was by this time completely under the supervision of the German Army Heeresgruppe Süd, consequently the entire country became a military area.

On the 6th January 1945, all of the Hungarian fighter units, namely the six squadrons of the *101. Vadászezred* – 101st Fighter Regiment – and the 102/1. and 102/2. Squadrons of the *102. Vadászosztály* came under control of the German fighter unit, Stab JG 76. This formation was in Hungary without aircraft of their own, but directed all German and Hungarian fighter force combat operations against the Soviet forces. In the first week of January 1945, the 102nd Fighter Group became a part of the 101st Fighter Regiment, as the third group i.e. *101/III. Vadászosztály*. The old 102nd first squadron became the 101/7, i.e. the seventh squadron of the 101st Fighter Regiment; the old 102/2. the new 101/8. Some days after this, the *Zöld Béka* – Green Frog – ferry pilot and transport squadron was recalled from the Western Front and also became a part of the 101st Regiment, being renumbered 101/9. as the ninth squadron. Previously, these pilots delivered German fighter

Below: *By the end of the war, the remnants of the MKHL and the Luftwaffe were scattered far and wide. These assorted trainers were all found at Klagenfurt in Austria. There are three Hungarian Arado Ar 96B's visible, G-935 being the nearest. Despite the late date, the replacement rudder still retains the chevron marking, obsolete in March 1942. Ezredes – Colonel – János Németh of the MKHL and the men under his command captured and held the airfield at Klagenfurt for the British in March 1945, for which he was given a commendation (Ketley)*

aircraft from the factories to the Western front, but did not take part in combat. Now, they compensated for the losses of the *101/III. Vadászosztály.*

From the 22nd January 1945, the whole *101. Vadászezred* – 101st Fighter Regiment – was subordinated to the *102. Repülodandár* – 102nd Air Brigade, thus becoming a 'front' i.e. tactical, air force.

On the 21st-22nd March 1945, the Soviet offensives across the River Danube smashed through the defences on the line of the Balaton-Velence lakes, near Esztergom, therefore the *101/I.* and *101/III. Vadászosztály* – 1st and 3rd Fighter Groups – withdrew from Veszprém and retreated to Kenyeri. On the 24th March they moved to the Western frontier and settled on the airfields of Söpte and Vát, but they were unable to commence operations because of the waterlogged ground conditions.

On the 27th March 1945, the *ezred* – Regiment – moved from Szombathely to Grosspetersdorf, then from the 29th March began another transfer to the airfield at Wiener-Neustadt, and within a couple of days the army and air units had effectively left the country. The battles in Hungary ended on the 16th April 1945, and not on the 4th as is often stated, while the air formations which remained in Austria continued the hopeless struggle from bases at Tulln and Eferding until the 30th April.

The last known victory of the *101. Vadászezred* – Fighter Regiment – was on the 17th April against a Yak-9 fighter. The last mission took place on the 30th April above Brno. All military operations then stopped because of the lack of fuel, and the collapse of Luftflotte 4.

Surviving members of the regiment set fire to their remaining aircraft and laid down their weapons at Eferding, shortly before the arrival of American troops on the 5th May, 1945.

Combat Record of the *101. Vadászezred 'Pumas'* – 101. Fighter Regiment – for 1944

No of Missions	175
Aircraft Deployments	1414
Victories Claimed	
1 engine	54
2 engine	35
4 engine	72
Total	161

Note:
The unit received 400 Messerschmitt Bf 109G-6, G-10 and G-14 fighters in this period. Only 80 were useable. 70 were under repair at the end of 1944

Combat Personnel Losses in 1944

Killed in Action	39
Killed by Bombing on Ground	13
Missing	3
Wounded	18

Total Combat Personnel Losses
January 1944–8th May 1945

Killed & Missing	67
Killed in Flying Accident	9
Wounded	25
Confirmed POW	8

Personnel Roster of the *Magyar Királyi Honvéd Légiérö* – Royal Hungarian Air Force

Year	Staff		Squadrons	Total Aircraft
	Officers	Men		
1939-40	393	3,273	32	444
1940-41	472	8,370	30	476
1941-42	561	15,121	19	323
1942-43	779	20,082	22	769
1943-44	945	24,290	29	864

In summer 1944 the maximum number of aircraft rose to some 900 because the Germans supplied a quantity,

Combat Record of the *101. Vadászezred 'Pumas'* – 101. Fighter Regiment – for 1945

No of Missions	260*
Aircraft Deployments	1696**
Victories Claimed*	
US	7
Soviet	144
Probables	10
Total	161

Notes:
* Data for February and May is missing
** Data for April is missing. The unit lost 19 aircraft totally destroyed

Combat Personnel Losses January and May 1945 – all other records are lost

Killed	7
Missing	1
Killed in Flying Accident	1
Wounded	1
Confirmed POW	2

Combat Record of the *102. Vadászosztály* – 102. Fighter Group – for 1944

No of Missions	1673
Hours over enemy territory	1955
Victories Claimed	85 (All Soviet)
Combat Personnel Losses in 1944	
Killed in Action	9
Aircraft	
Damaged beyond repair	19
Damaged, but repairable	51

The Reckoning

At the end of 1944 a despatch from the 8 Jagddivision in Vienna arrived at the *101. Vadászezred* – 101st Fighter Regiment – Headquarters concerning the results of the period 1st January 1944-30th November 1944. As the Hungarian estimate was not ready for some reason, they used this one in the absence of anything more precise. This despatch was probably an extract from the 8 Jagddivision *Kriegstagebuch*, which has been missing ever since then. According to this, during the course of 1,581 fighter and bomber missions they shot down 114 four-engined bombers and 66 single- and twin-engined fighters, altogether 180 aircraft. i.e. for every 8.8 missions there was 1 victory. What this effectively means is that, on average, the Hungarian forces deployed 12 aircraft 54 times, and claimed 107 enemy aircraft shot down for the loss of 78 of their own, whereas the Germans deployed 12 aircraft 77.6 times, shot down 73 and lost 88. This means that the Hungarian missions were more effective even when they were not as strong as the Germans.

If the data for the period of 30th November 1944-30th March 1945 is taken into account as well, in three missions by Hungarians, 17 fighters flew against American aircraft, and shot down 3 bombers and one fighter. Combining this information concerning the *101. Vadászezred* in the period of 1st May 1944-30th November 1944 and taking into account the results of the 2/1., 1/1., and 5/1. squadrons, the Hungarian fighter forces in the period of 1st January 1944-30th March 1945 sent 795 aircraft on 66 missions and achieved 129 air victories. In other words, for every 6.16 missions, one enemy aircraft was shot down, which is no worse than the results of the Luftwaffe. It is difficult to compare this with the American or British results as their methods of verifying and evaluating their achievements were different.

The Fighter Aces

Available surviving data is difficult to summarise. There is no trace of an official list of results from the then Hungarian Airforces Command. The achievements of individual pilots were kept by their own squadrons and when they were posted to another unit or recommended for decoration then their achievements were mentioned in their documents. As most of them

Combat Records of the *2/1. & 1/1. Vadászszázad, 101. Vadászosztály* & *101 Vadászezred*
January 1944 to March 1945

2/1. Vadászszázad (Ferihegy)

1944	No of Missions	Aircraft Deployed	Victories (Claimed/Confirmed) 1 engine	2 engine	4 engine	Total	Combat Personnel Losses Killed	Wounded
January	2	23	-	-	-	-	-	-
February	1	12	-	-	-	-	-	-
March	1	11	-	-	-	-	2	-
April	8	104	-	-	5/3	5/3	-	1

1/1. Vadászszázad (Szolnok)

1944	No of Missions	Aircraft Deployed	Victories (Claimed/Confirmed) 1 engine	2 engine	4 engine	Total	Combat Personnel Losses Killed	Wounded
April	2	10	-	-	2/0	2	2	-
TOTAL	14	160	0	0	7/3	7/3	4	1

101. Vadászosztály (Veszprém)

1944	No of Missions	Aircraft Deployed	Victories (Claimed/Confirmed) 1 engine	2 engine	4 engine	Total	Combat Personnel Losses Killed	Wounded
May	2	45	1/1	-	4/4	5/5	1	3
June	14	196	5/2	26/11	14/8	45/21	9	3
July	19	196	6/4	6/2	26/21	38/27	4	3
August	9	121	3/0	1/0	12/1	16/1	4	1
September	1	2	-	-	-	-	-	-
TOTAL	45	560	15/7	33/13	56/34	104/54	18	10

101. Vadászezred (Veszprém – Kenyeri)

1944	No of Missions	Aircraft Deployed	Victories (Claimed/Confirmed) 1 engine	2 engine	4 engine	Total	Combat Personnel Losses Killed	Wounded
October	1	8	2/2	-	-	2/2	2	1
November	8	80	1/1	1/0	10/0	12/1	7	2
December	-	-	-	-	1/0	1/0	-	-
1945								
January	-	-	-	-	-	-	-	-
February	1	6	-	-	1/0	1/0	-	-
March	2	11	1/0	-	2/0	3/0	-	-
TOTAL	12	105	4/3	1/0	14/0	19/3	9	3
TOTAL	71	825	19/10	34/13	77/37	130/60	31	14

were annihilated, we can only trace secondary sources. In doing so we have had to look into contemporary papers and books. A detailed summary with all the names can only be a part of a later, larger publication. Until then we follow the Hungarian screening system, which only recognized total or shared air kills, in the latter case the victory was to be mentioned by all pilots involved. This, of course, could have been the source of later errors.

Research indicates that the following further victories were scored, although the data is incomplete:

3 × pilots or crews	4 victories claimed
16 × pilots or crews	3 victories claimed
24 × pilots or crews	2 victories claimed
48 × pilots or crews	1 victory claimed

In total 125 men claimed 513 confirmed air victories, plus 21 unconfirmed, or only possibles. Research continues, and it seems certain that the number of pilots who reached 5 air victories and 'ace' status is larger.

Hungarian fighter aircraft did not carry gun cameras, therefore the results were verified by the wingmen or unit crew members afterwards. Having correctly recorded the areas where the combat took place, confirmation was later requested from witnesses in the area and from their own units. This could be done over their own territory, not behind the front line. Accordingly, from 1939 until the autumn of 1942 and then from February until September 1944, the results and losses of the air battles were known in great detail. From the autumn of 1942 on the Eastern Front, most combats happened behind the Soviet lines, therefore the results of the air battles were untraceable many times, verification being out of the question. These results were mentioned as possible air victories in some of the despatches, but in many units, these claims were not mentioned at all. Combats against enemy units by the Hungarian fighter forces mostly began at a height of 7-8,000 metres, usually finishing near ground level in order to break off from the attacks. During these activities many aircraft on both sides came back damaged, but returned to their own bases, though they were claimed to have been shot down. On the other hand, many damaged enemy, particularly American, aircraft flew out of Hungarian air space, to later crash in Croatia or in occupied Yugoslavia, or in the Adriatic sea. From these facts we can safely say that there is a numerical difference between the 'confirmed' victories and those 'confirmed shot down'. In Hungary, the German radar units, which were subordinated to the 8 Jagddivision could watch the Adriatic for enemy aircraft until September 1944, after this they were unable to do so.

The MKHL Headquarters evaluated the Hungarian fighter defences on the basis of the facts which they requested from the Germans, as illustrated by the following extract from the records of the *101. Vadászezred* – 101st Fighter Regiment – in the Hungarian War History Archives:

"At the end of the summer the HQ of the Hungarian Air Forces asked for data from the German Jagdabschnittführer about the achievements of the Hungarian and German fighters above Hungarian territories, and here they are:

101. Vadászezred:

Number of aircraft sorties:	*649*
Aircraft shot down:	
With 4 engines	*61*
With 2 engines	*32*
With 1 engine	*14*
Total	*107 aircraft*
Losses: 30 pilots in 78 aircraft	

Luftwaffe day fighters:

Number of aircraft sorties:	*932*
Aircraft shot down:	
with 4 engines	*53*
with 1-2 engines	*20*
Total	*73 aircraft*
Losses: 43 pilots in 88 aircraft	

1 air victory for every 6 Hungarian missions
1 air victory for every 12.8 German missions

These are the evaluations of the data above, but the numbers speak for themselves. The German air victories were 68% of the Hungarian air victories. The number of aircraft available to the Hungarians was 70% of the Germans.

The Hungarian fighters achieved the above results despite their more modest means, technical knowledge, fewer flying opportunities, more modest levels of training, lack of experience and with generally smaller fighting forces than the Germans, and even though they were altogether in a more disadvantageous situation with regard to the enemy."

Hungarian Aces

The pilots of the *Magyar Királyi Honvéd Légierö* who scored 5 or more victories in aerial combat in WW II

Name & Rank	Unit	Missions	Claims	Notes
Zászlós Dezső Szentgyörgyi*	101/2. V. Szd	218	32 + 2	KIFA 1971
Főhadnagy György Debrődy	101/3. V. Szd.	204	26 + 2	Wounded 16.9.1944. Did not fly again
Hadnagy László Molnár**	101/3. V. Szd.	132	25	KIA 7.8.1944
Hadnagy Lajos Tóth***	101/2. V.Szd.	180	24 + 1?	Executed 11.6.1951
Hadnagy Miklós Kenyeres	102/1. V. Szd.	?	18 + 1	Soviet POW 6.2.1944
Őrmester István Fábián****	101/2. V. Szd.	120	15.5 + 1	
Százados László Pottyondy	101/8. V. Szd.	?	13 + 1	
Főhadnagy Ferenc Málnássy	101/III Oszt.	?	13 + 2	KIA 13.3.1945 101/8. V. Szd.
Alezredes Aladár Heppes	101. V. Oszt.	?	8 + 4	CO 101 Fighter Regiment*****
Hadnagy Kálmán Nánási	101/5. V. Szd.	?	12	
Zászlós István Kálmán	101/III Oszt.	?	12	
Hadnagy József Málik	101/1. V. Szd.	60	11 + 1	KIA 16.4.1945
Főhadnagy László Dániel	101/3. V. Szd.	58	11.75	
Őrmester Lajos Budai	101/1. V. Szd	33	9	KIA 18.3.1945
Főhadnagy Pál Irányi	101/1. V. Szd.	?	8	Until 3.3.1945
Főhadnagy Endre Német	1/1. Tf. Szd	59	8	Pilot of a fighter-reconnaissance aircraft
Hadnagy Kálmán Szeverényi	101/7 V. Szd.	?	7	KIA 22.2.1945
Hadnagy György Michna	101/1. V. Szd.	?	7	
Százados György Bánlaky	101/4. V. Szd.	?	7	KIA 6.11.1944
Szakaszvezető László Pál Szikora	101/1. V. Szd.	?	7 + 1?	Wounded on or about 3.3.1945
Hadnagy Imre Páncél	1/1. Vegyes V. Szd.	?	6 + 1?	MIA 11.1.1943 First MKHL ace
Százados József Bejczy	101/3. V. Szd.	?	6 + 1?	MIA 4.11.1944
Őrmester Mátyás Lőrincz	101/2. V. Szd.	?	6	KIA 5.11.1944
Főhadnagy Tibor Papp	101/6. V. Szd.	62	6	
Őrmester János Mátyás	101/3. V. Szd.	?	5 + 1?	
Főhadnagy Ervin Flóznik	101/4. V. Szd.	?	5	MIA 15.1.1945
Főhadnagy Béla Füleky	101/4. V. Szd.	?	5	MIA 4.4.1945
Hadnagy Pál Forró	101/2. V. Szd.	?	5	
Hadnagy Tibor Tobak	101/3. V. Szd.	?	5	Wounded 16.4.1945
Hadnagy Sándor Hautzinger	5/2. V. Szd.	?	5	Wounded 9.1943 Did not fly again
Hadnagy Mihály Karátsonyi	101/3. V. Szd.	?	5	Wounded 7.8.1944
Őrmester Pál Kovács	101/3. V. Szd.	?	5	KIA 22.8.1944
Százados István Ujszászy	101/2. V. Szd.	?	5	
Százados Antal Nagyernyei	1/1. Tf. Szd.	?	4 + 1?	1941-42 only. Pilot of a reconnaissance aircraft

As the MKHL was a part of the army for most of its existence, ranks were those used in the army and consequently are more akin to those of the USAAF than the RAF:

Szakaszvezető	Lance Sergeant	*Főhadnagy*	Lieutenant	*Vezérőrnagy*	Major General
Őrmester	Sergeant	*Százados*	Captain	*Altábornagy*	Lieutenant General
Törzsőrmester	Staff Sergeant	*Őrnagy*	Major	*Vezérezredes*	General
Zászlos	Cornet or Ensign	*Alezredes*	Lieutenant Colonel	*Tábornagy*	Field Marshal
Hadnagy	Second Lieutenant	*Ezredes*	Colonel		

* In the latter part of his service Szentgyörgyi was an *Hadnagy-helyettes* – Deputy Second Lieutenant, the highest rank attainable by an NCO in the MKHL. Records differ as to the true total of his victories which range from 29.5 to 34. Two others were German aircraft shot down in error, and were not included

** If Molnár's claims are tabulated following American methods he would have 23.25.

*** Tóth claimed 4 more on the Eastern Front in 1943-44 for which there is no documentation

**** *Törzsőrmester* in 1945, with 13 more unconfirmed claims

***** The four 'probables' claimed by Aladár Heppes in 1943 over the Eastern Front were achieved during a combat without a wingman, therefore although these are mentioned in the summary they could not be confirmed

This list is not complete as other crew members also had air victories. Many units left behind only total results, not listing individuals

CAMOUFLAGE AND MARKINGS

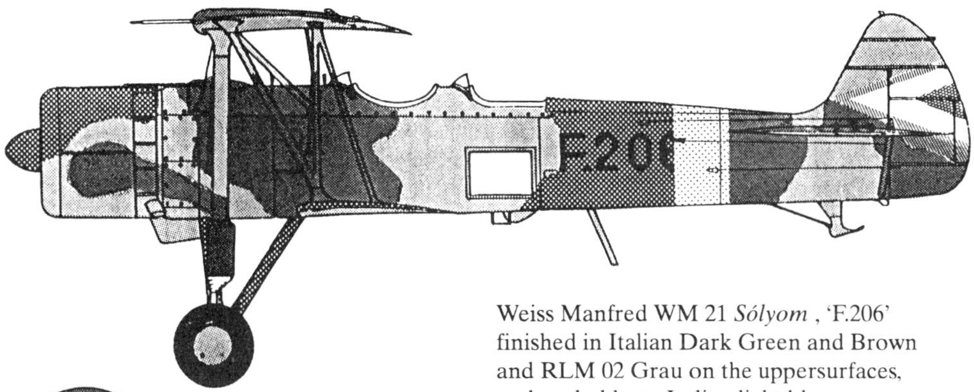

Weiss Manfred WM 21 *Sólyom* , 'F.206' finished in Italian Dark Green and Brown and RLM 02 Grau on the uppersurfaces, and probably an Italian light blue-grey below. Sole theatre marking is a yellow fuselage band

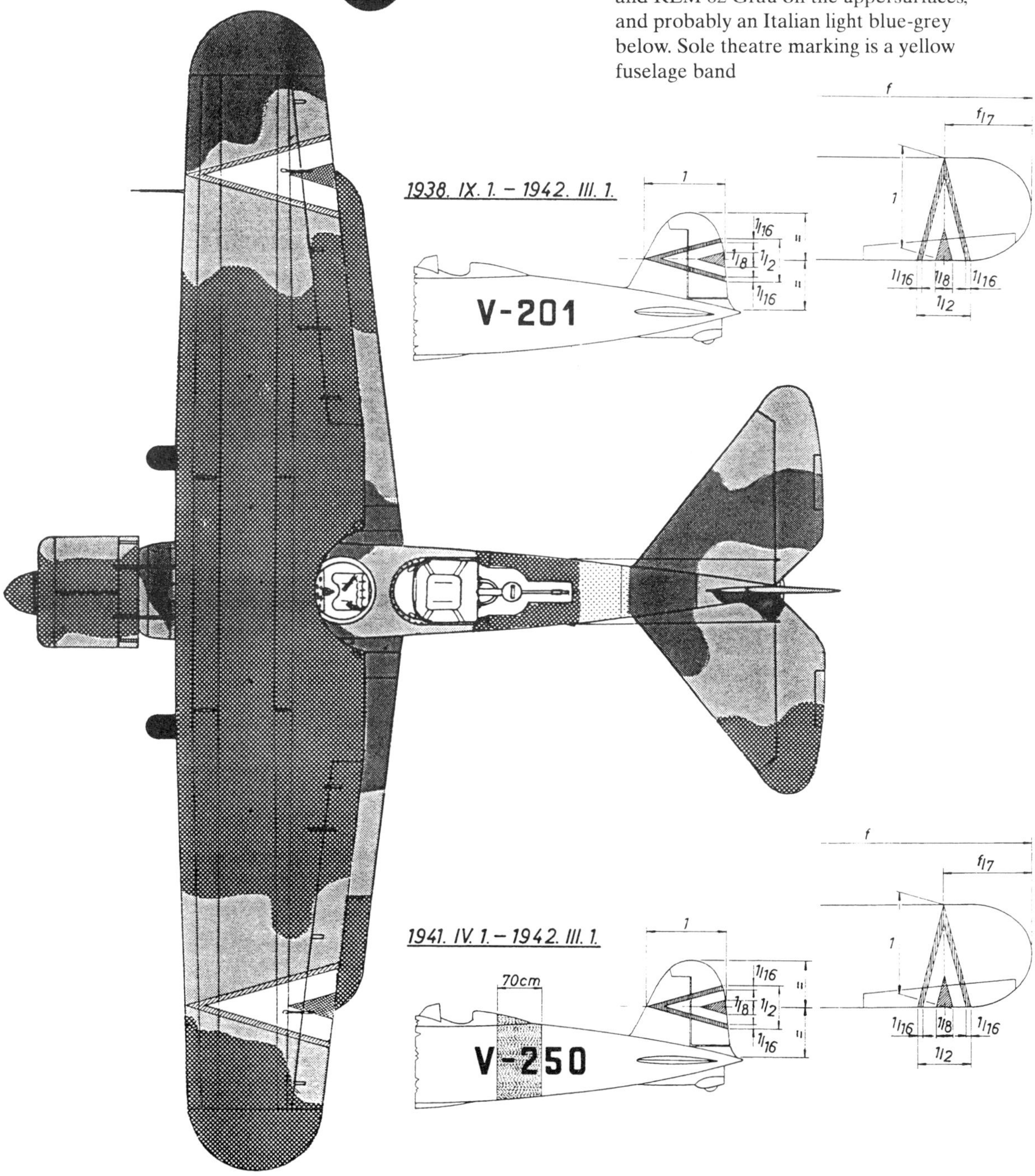

1/72nd SCALE

CAMOUFLAGE AND MARKINGS
National Insignia

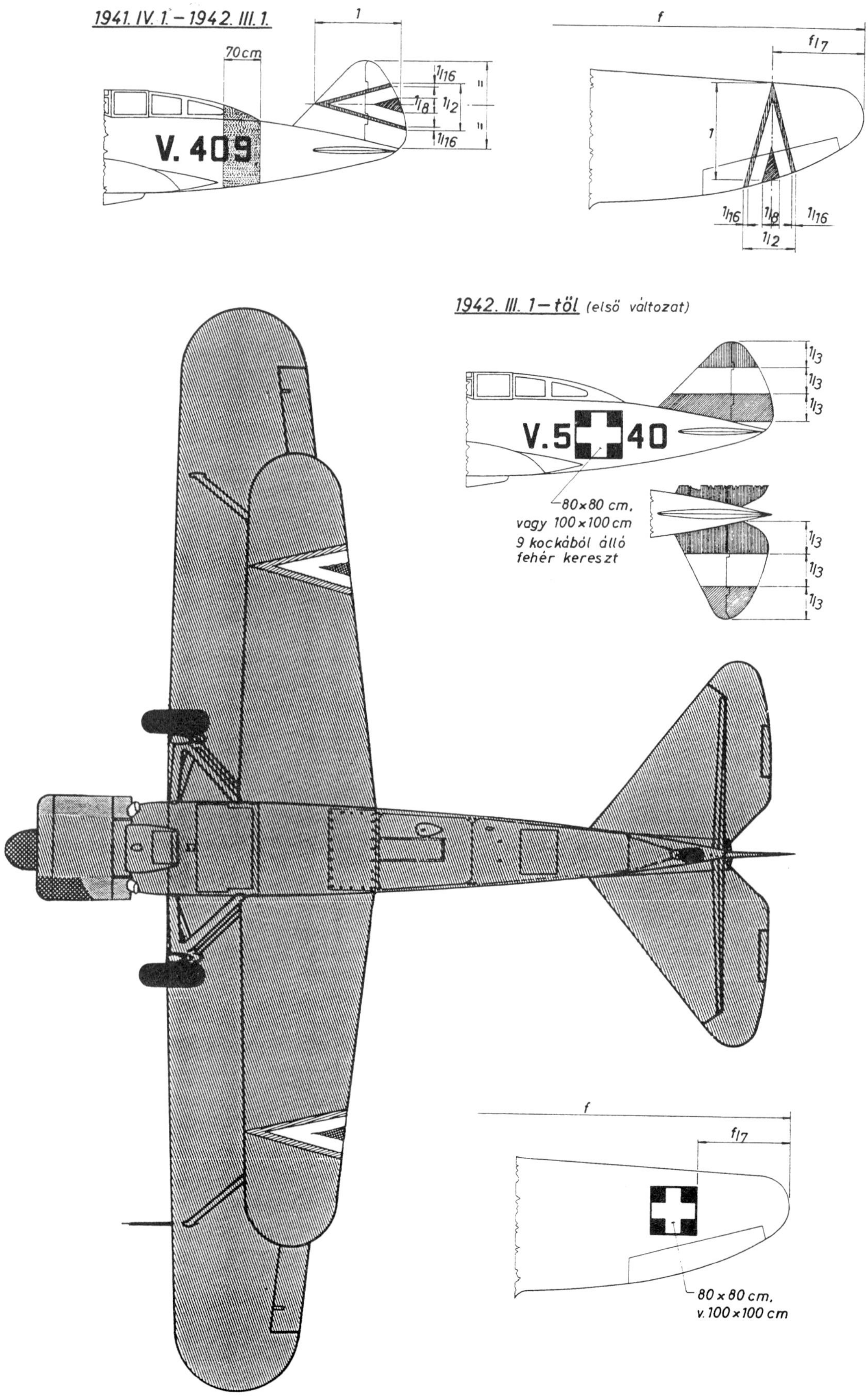

CAMOUFLAGE AND MARKINGS

National Insignia

1942. III. 1. – 1944 öszéig

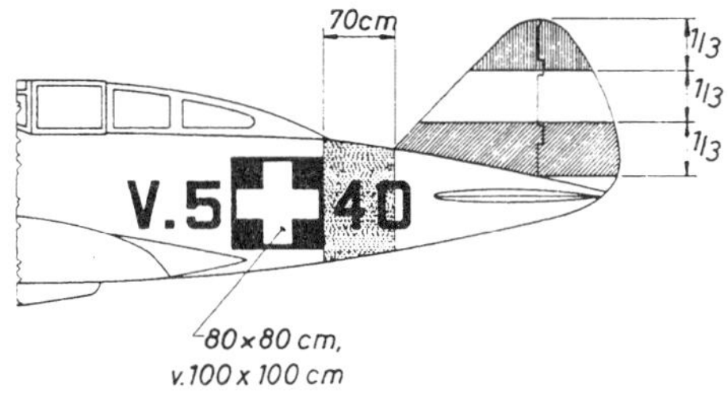

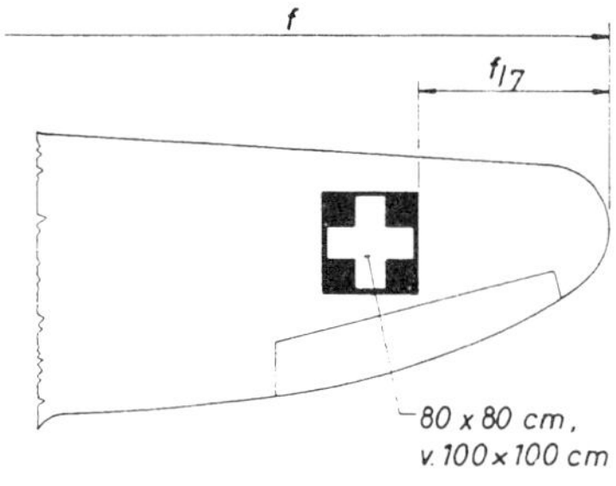

1942. XI. – 1943. II.

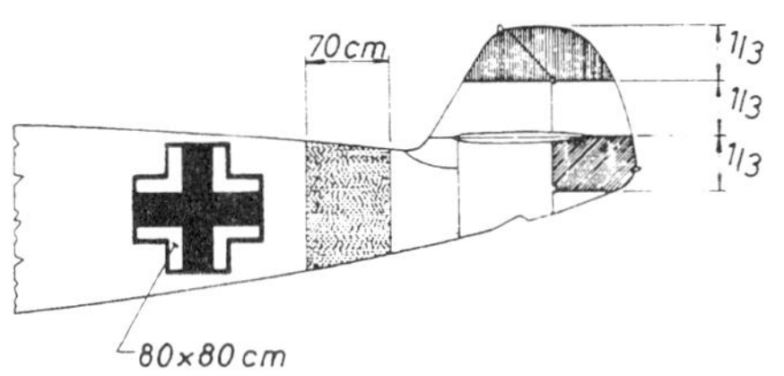

(Csak a keleti fronton lévő Bf-109 F4 vadászgépeken)

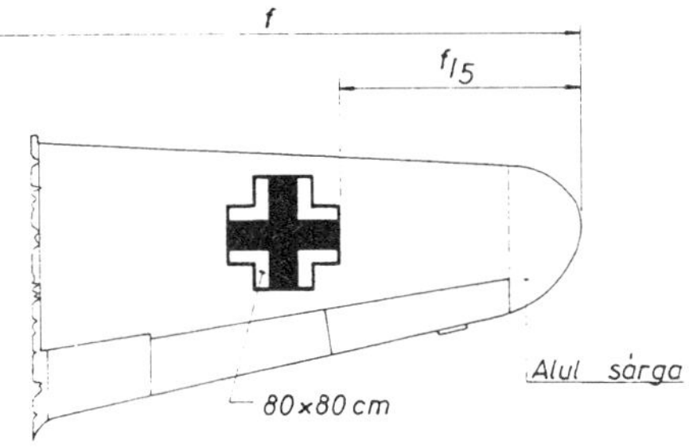

1943. II-től 1944 nyaráig

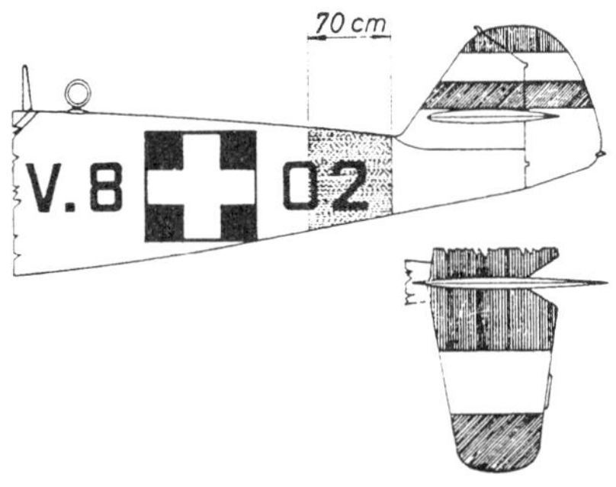

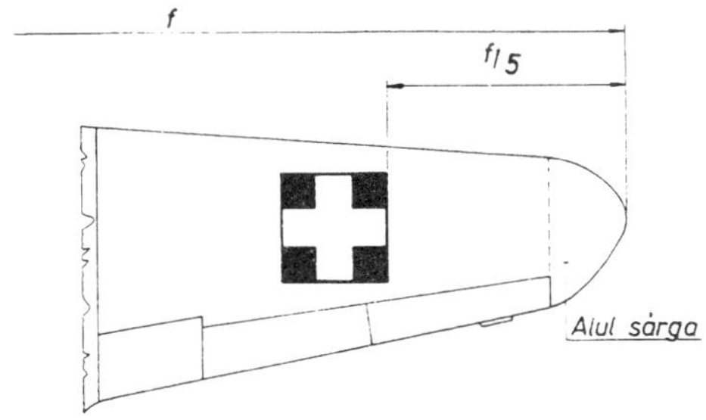

1944. IX. – 1945. V.

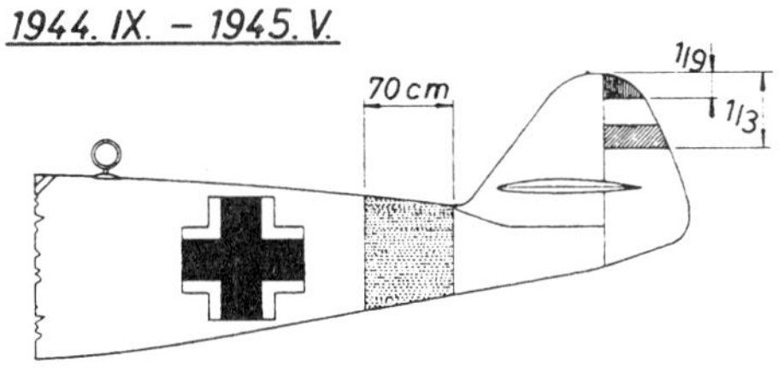

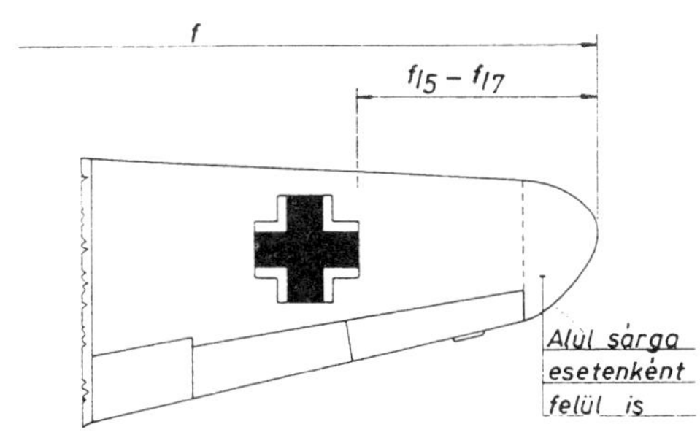

1944. IX-töl – 1945. V.

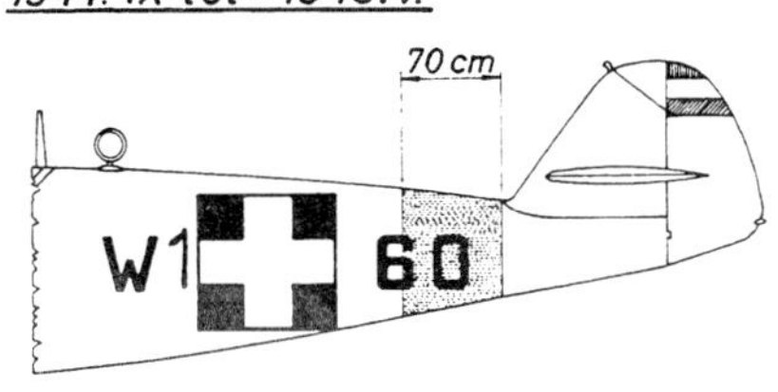

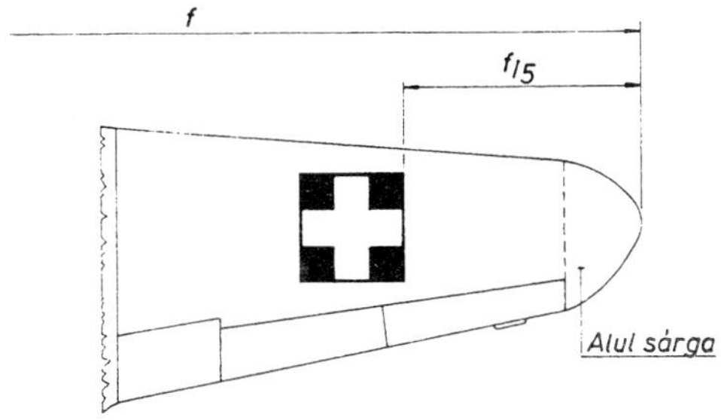

Messerschmitt Me 210Ca-1 Maintenance Markings

Based on official Dunai Aircraft Factory drawings 8-210 Ca-1

Canopy & Access Inscriptions

Position	Qty	Size in mm	
1	1	6	Ülésszabalyzó
2	1	5	Beszállo létra ki-be működtető
3	1	6	Önműködő távkapcsoló
4	1	5	Futóművet 280 km/h felett kiengedni tilos!
5	1	8	Magassági légző-szükségkioldó
6	1	8	Üresjárat zuhanórepüléskor
7	1	5	Rögzítve - farokkerék - oldva
8	1	6	A szilárd antennával hosszú hullámon tűzveszély miatt tilos rádiózni!
9	4	-	A töltények és hevederek helyzetének kijelzője
10	1	4	Biztosan kiengedhető üzemanyag mennyisége --- liter

N-210 Ca.1-BL 10.21.			Fuselage inscriptions
11	1	16	Hálózati külső csatlakozó 24 Voltos
12	6	10	Gyártási szám
13	2	10	Elreteszelni
14	2	20	Felbakolás előtt 500 kg-al megterhelni
15	2	20	Tömlő nyomás 3,5 atü
16	2	20	Villás emelővel itt emeld
17	1	10	Beszállo belépő keretezve
18	4	14	Itt fogózkodj, keretezve
19	1	20	Oxigén! Olajtól, zsirtól óvni/ jelölő rajzzal
20	1	-	Mentőcsomag helye, vörös ke reszttel jelölni
21	2	10	Jelzés a vízszintes vezérsik 0 állásához
22	2	10	Csak itt fogd meg!
23	2	30	Szigetelőlakkal bevonva!
24	1	40	1-9, 1-9 Keret megjelölés
25	2	10	A burkolat levétele előtt a gyujtásáram kapcsolót kikapcsolni! A burkolat felrakása utan a gyujtásáram kapcsolót bekapcsolni!
26	2	20	A csószóbakkal itt támaszd meg!
27	2	20	Orsós emelővel itt emeld!
18	1	12	/alá/ Indulas előtt: A vizszintes vezérsik 0 állására ügyelj. A zuhano fék hidraulikus kapcsolóját nyomd "Be" helyzetbe
29	-	-	
30	-	-	

8.210. Ca-1 Bl. 8,9 lap			Inscriptions above the wings
31	6	25	Sárga háromszögben: 87
32	2	10	Barna háromszögben: Aeroschell töltő
33	2	16	Mérőmutatót figyeld!
34	2	16	A rögzítőcsavarok oldásakor a bizto sítást csőkulccsal nyomd le!
35	2	16	Ne fogd mg!
36	2	25	Csak itt lehet alatamasztani!
37	1	20	AB 11 Présolaj
38	1	8	Olajat csak szűrőn keresztül betölteni
39	2	-	A kormánylap rögzítő ollók felrakási helyének jelölése
40	6	6	Tartály űrtartalma ... liter

8-210. Ca-1 Bl 8,9 lap			Inscriptions under the wings
41	1	20	Felbakolás előtt a törzsvéget 500 kg- al megterhelni!
42	4	16	A fedél levétele előtt a tartályt üritsd ki!
43	2	20	A két fedél levétele előtt a motort támaszd alá!
44	1	16	Préslevegő a futóművesz kioldáshoz és a géppuskához 150 atü/jel
45	2	16	Itt feltámasztani!
46	2	10	Gyorsleállító - Téli indítás!
47	1	20	A hydraulika ellenörző csatlakozása
48	2	30	Szigetelő lakkal bevonva
49	1	10	Közdarab az ellenőrző nyomásmérőhöz a II, narancs szersz.készlet használatához
50	2	6	A motor cserélhető, Messerschmitt A.G. Augsburg

8-210. Ca-Bi 11. lap			Engine Cowling Inscriptions
51	2	20	Kinyitni tilos!
52	2	20	Óvatosan nyisd! A hűtő a burkolatba van beépítve!
53	2	10	Vigyázz! Gőz!
54	2	6	A moto cserélhető, Messerschmitt A.G. Augsburg
55	-	-	

			Undercarriage Inscriptions
56	2	6-4	Vigyázz! A vontató kábelkötél minimum 10 méter legyen!
57	2	6-4	A Me-210 repülogep rugóstag terheletlenül 33 atü, kerékabroncsok 3.5 atü préslevegővel töltendők
58	2	4	Itt préslevegővel 80 atü-re utántölteni
59	2	3	Rugóstag nyomás terheletlenül 33 atü, olajnívó
60	-	-	
61	-	-	

			Flight Control Inscriptions
62	4	4	A kormánylap rögzítők felszerelési hely jelölése
63	2	-	A beállító jelek megjelölése!

NOTE: No official style for the lettering was specified, although it presumably followed the German pattern, but as the stencils were hand-cut there was probably some variation in style

Messerschmitt Me 210Ca-1 Maintenance Markings

Uppersurface marking locations

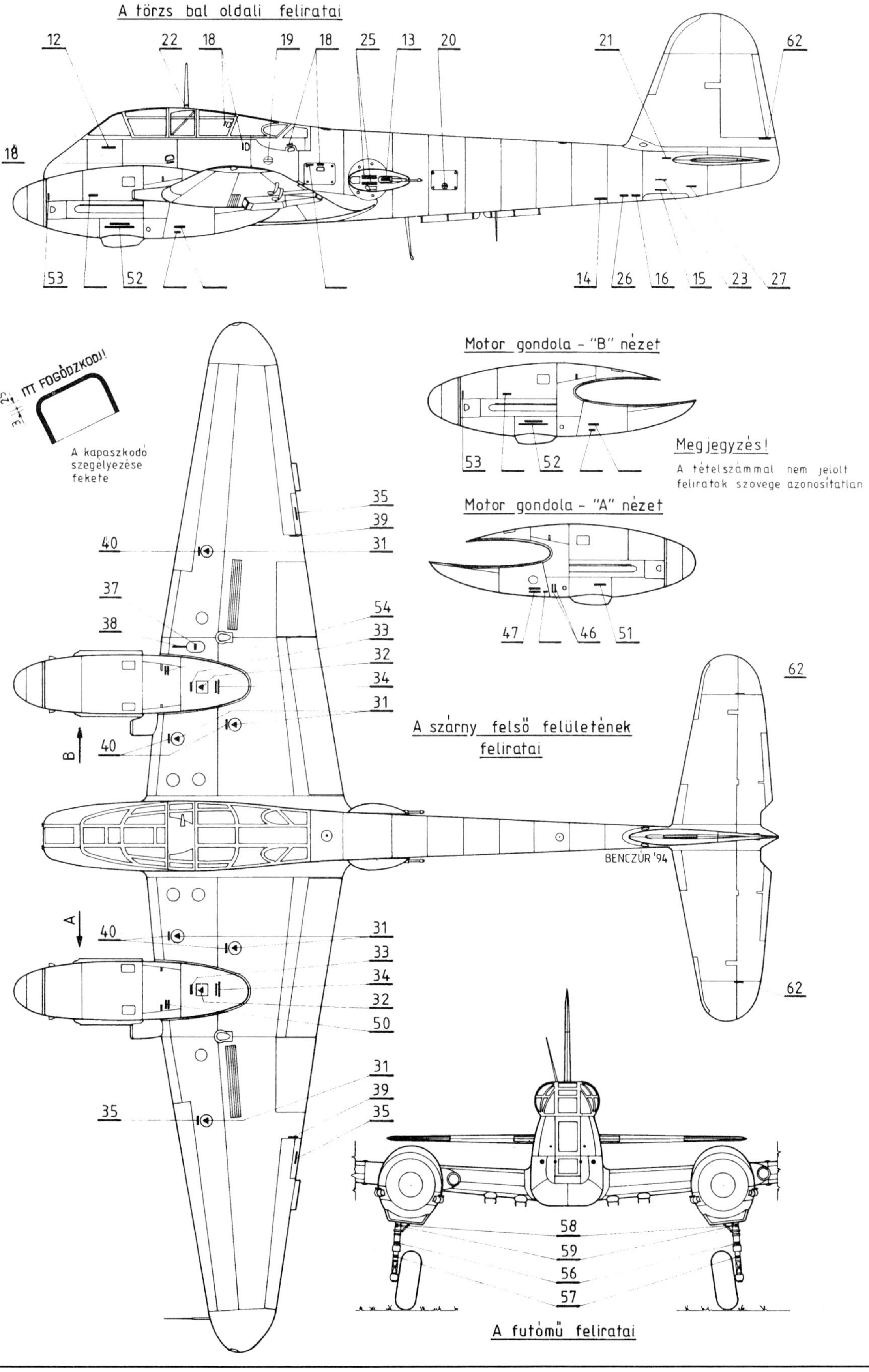

Messerschmitt Me 210Ca-1 Maintenance Markings

Undersurface marking locations

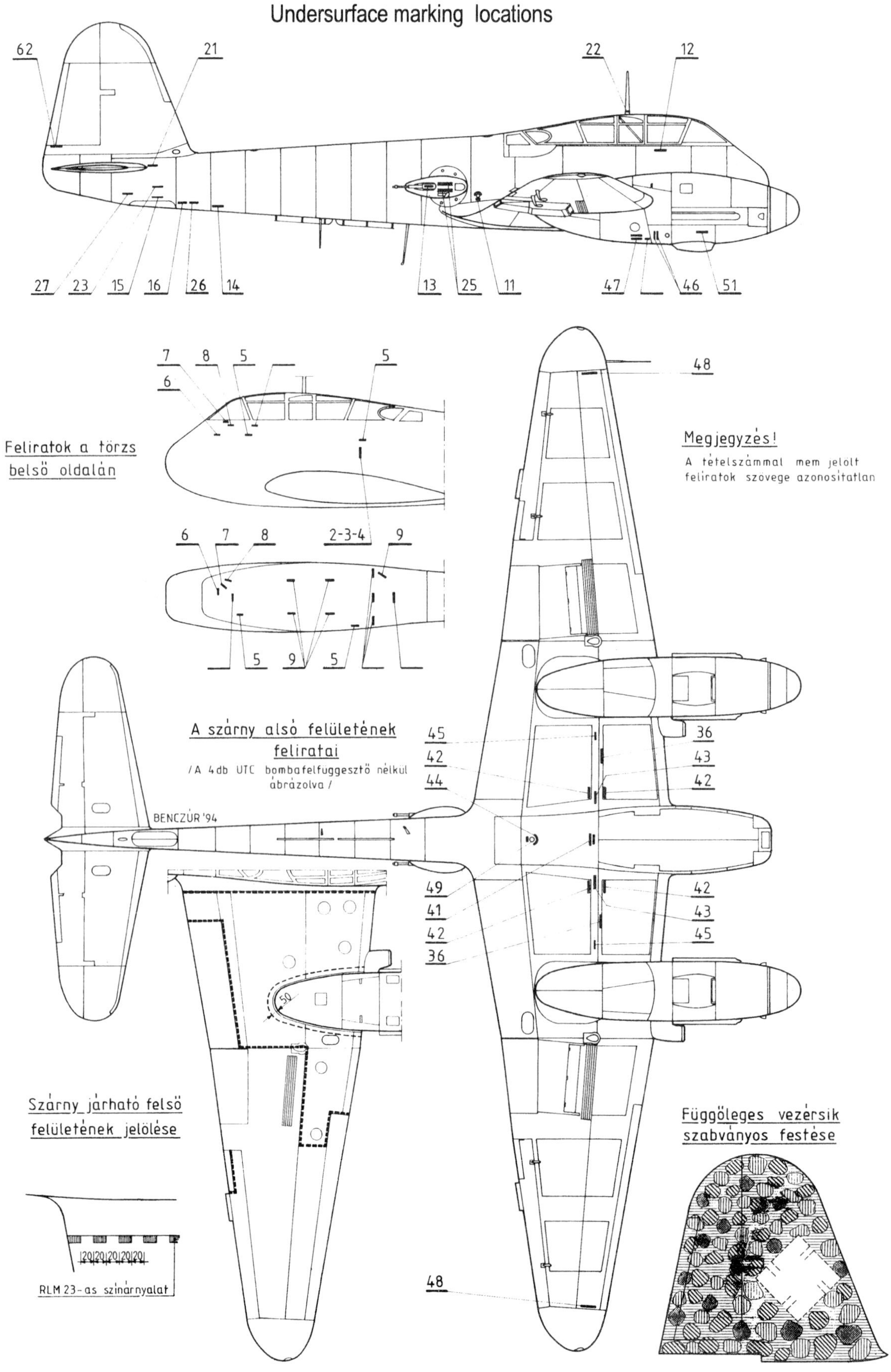

CAMOUFLAGE AND MARKINGS

Arado Ar 96A , G.4+48, in 1942 markings. RLM 70 Dunkelgrün above, RLM 65 Hellblau below. Yellow fuselage band

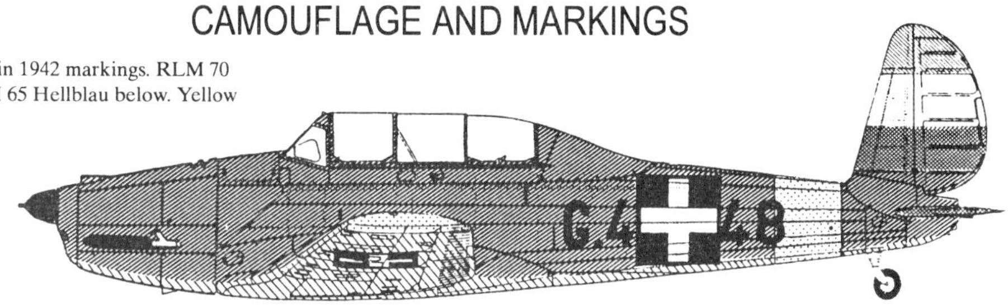

Arado Ar 96B , G.5+36, during 1943. Overall RLM 02 Grau with yellow fuselage band and underside wingtips

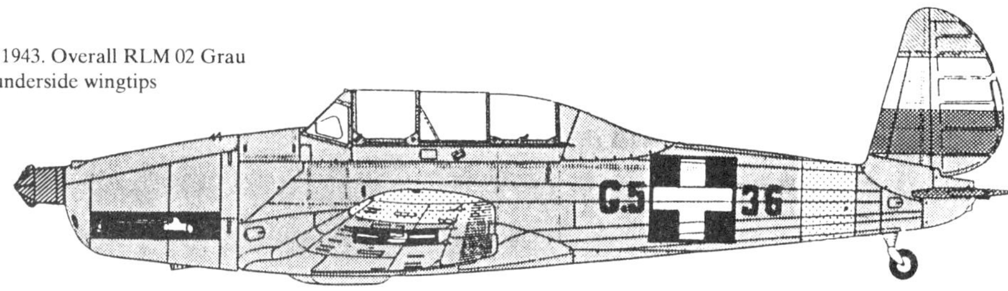

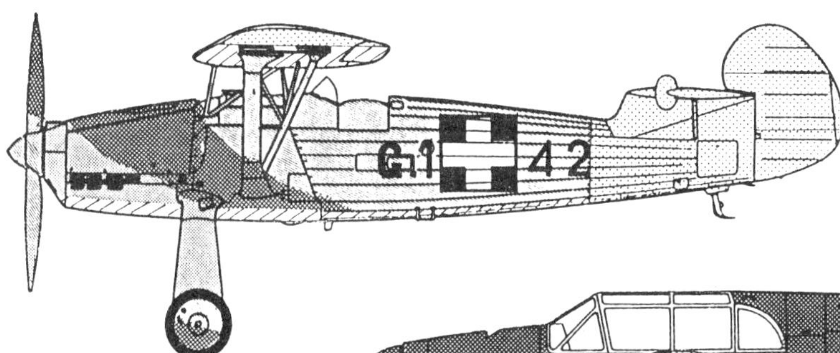

Focke-Wulf Fw 56 trainer, G.1+42, location unknown. Camouflage is probably a mixture of Italian light brown and dark green uppersurfaces, with light grey undersides. Underside wingtips and

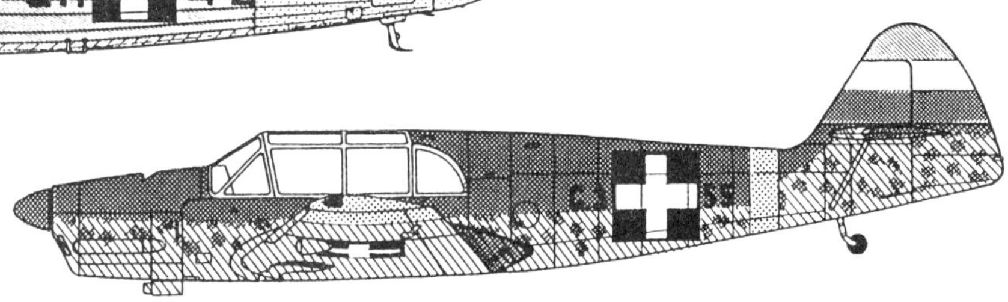

Messerschmitt Bf 108 serving with the *101. Vadászrepülő osztaly* – Fighter Group – as a courier and hack at Veszprém in 1944. Finish probably RLM 71/65 with 02 in the mottle. Yellow Eastern Front markings

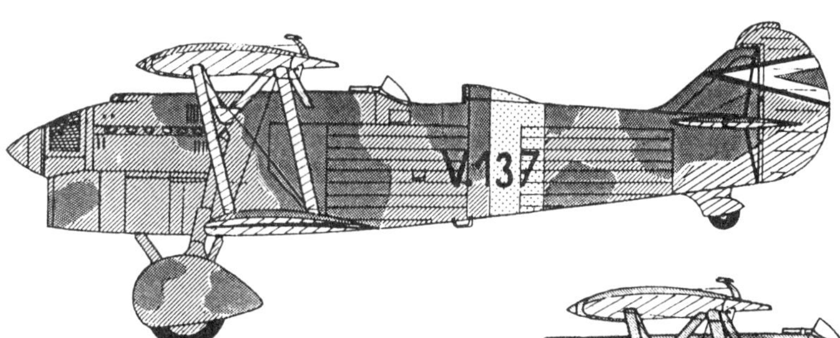

Fiat CR 32, V.137, of the *2/3. Vadászszázad*. Position of the yellow fuselage band over the Italian camouflage suggests that the date is early summer 1941

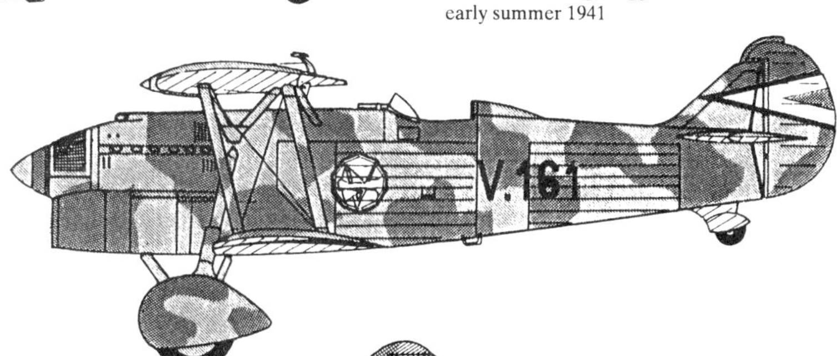

Fiat CR 32 of the *1/1. 'Ijász' Vadászszázad* – the Archers. V.161 wears standard Italian colours in one of the less common patterns. Börgönd, 1940

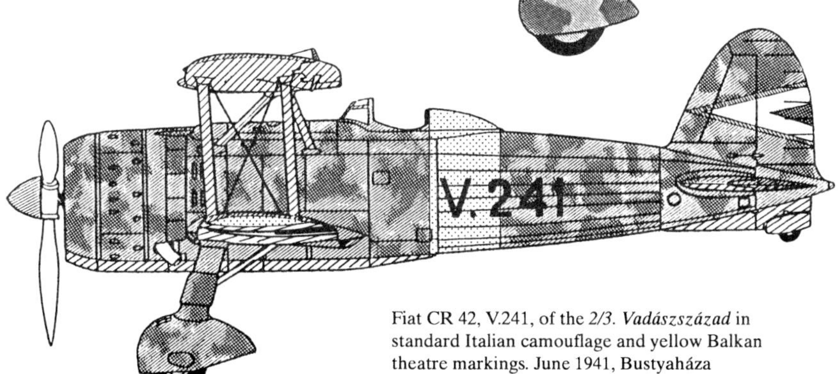

Fiat CR 42, V.241, of the *2/3. Vadászszázad* in standard Italian camouflage and yellow Balkan theatre markings. June 1941, Bustyaháza

1/72nd SCALE

CAMOUFLAGE AND MARKINGS

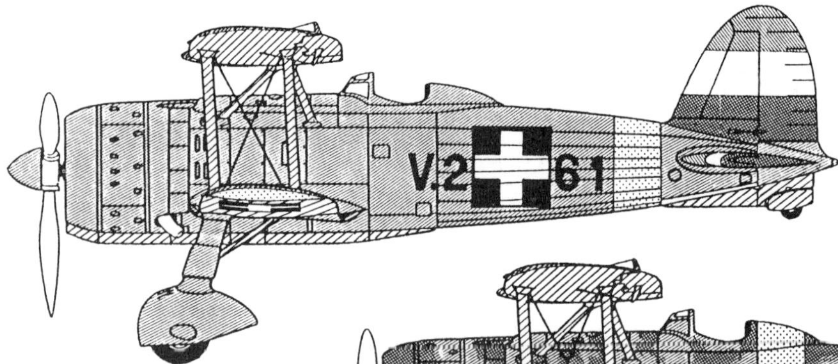

Fiat CR 42, V.2+61, of the *5/1. Vadászszázad* in 1943. This unit later became the *5/1. Éjszaki Vadászszázad* – Nightfighter Squadron – defending Budapest in 1944. Finish appears to be all-green with blue-grey undersides and yellow fuselage band

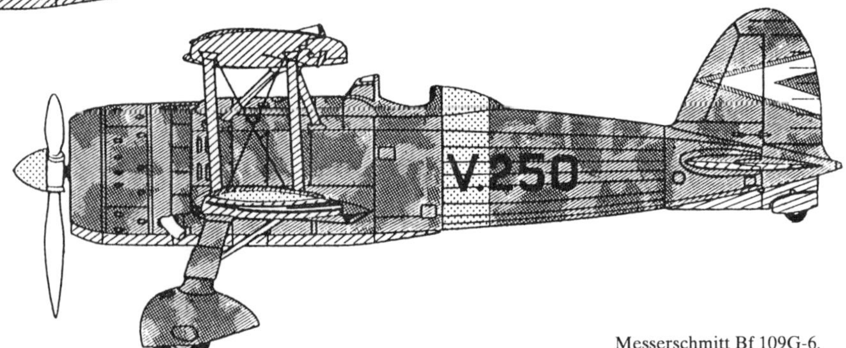

V.250, a Fiat CR 42 of *1/3. Vadászszázad*, on the Soviet Front, August 1941. Standard Italian scheme with yellow theatre markings

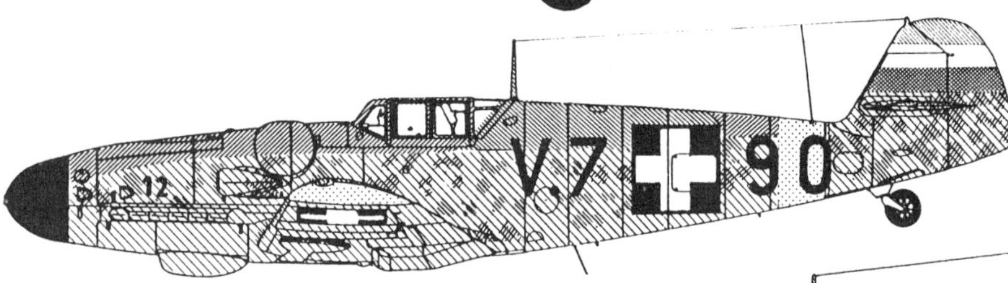

Messerschmitt Bf 109G-6, V.7+90, 3rd flight? of *2/1. Vadászszázad*. Budapest, Ferihegy, March 1944. Pilot *Tizedes* – Corporal – István Kis-Benedek. Red '12' just above exhausts identifies plane in squadron

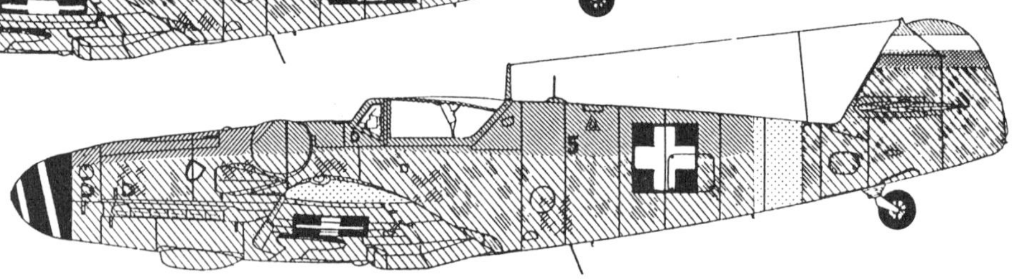

Messerschmitt Bf 109G-6, of the 102/1. Vadászszázad, Budapest-Budaőrs, 1944. Small red '5' just behind cockpit. 74/75/76 finish

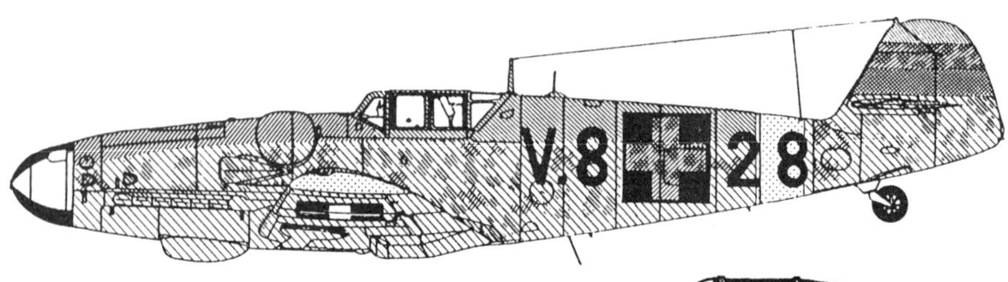

Messerschmitt Bf 109G-6, V.8+28, of the *102/1. Vadászszázad*, Veszprém, summer 1944. 74/75/76 finish. Note the 'hollow' crosses

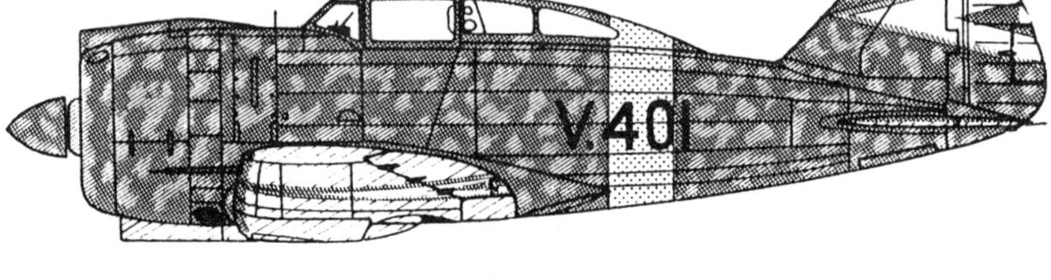

Reggiane Re 2000 *Héja* of the *1/2. Vadászszázad*, V.401, flown by the squadron commander, *Szazados* – Captain – János Gyenes. Soviet Front, August 1941. Typical Italian camouflage with yellow theatre markings

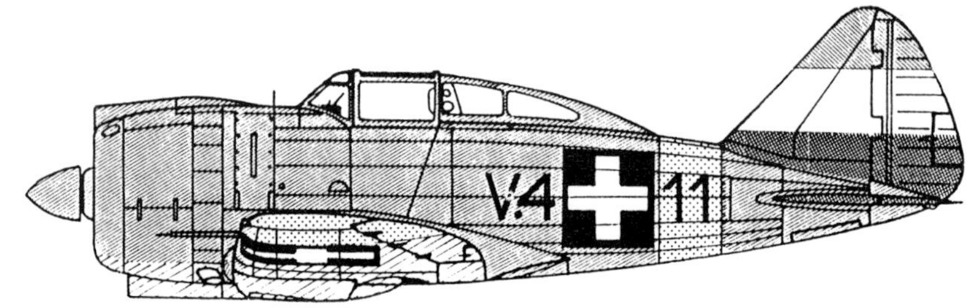

Reggiane Re 2000 *Héja* of the *1/1. Vadászszázad*, V.411, flown by the group commander, *Alezredes* – HQ Staff Lieutenant Colonel – Kalmán Csukas. Soviet Front, August 1942

1/72nd SCALE

CAMOUFLAGE AND MARKINGS

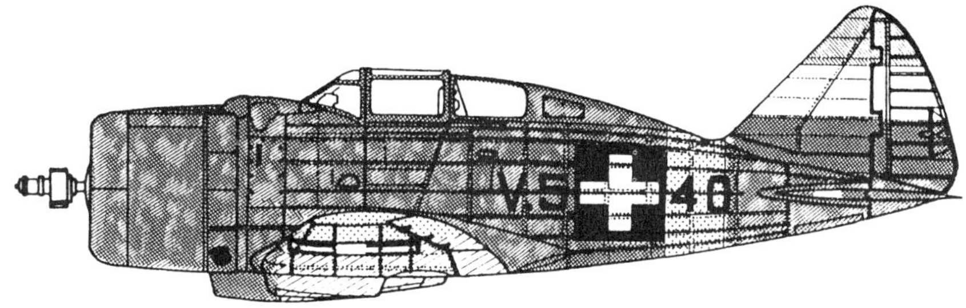

MÁVAG built *Hejá*, V.5+40, in 1944 at Szolnok, wearing Italian-style mottle uppersurfaces, light blue-grey below

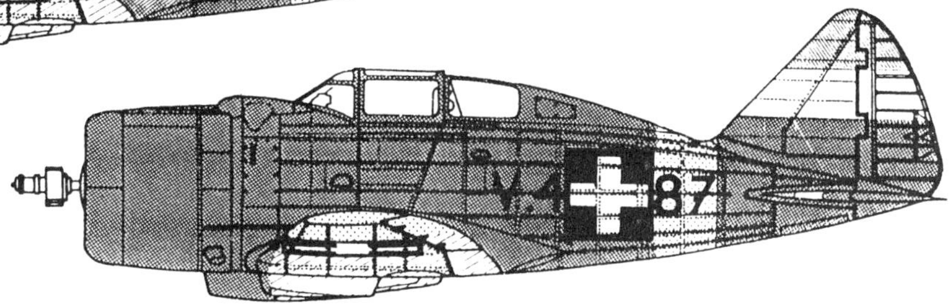

MÁVAG-built *Héja-M*, V.487, in 1943 at Budapest-Mátyásföld wearing solid dark green uppersurfaces, light blue-grey below

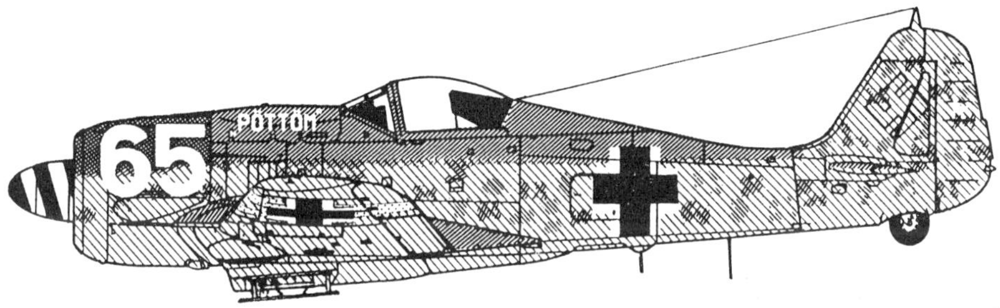

Fw 190F-8, B.5+65, of the *102/1. Vadászbombázószázad* in February-March 1945 at Kemenesszentpéter, Hungary. Standard Luftwaffe 74/75/76 finish, white number and name

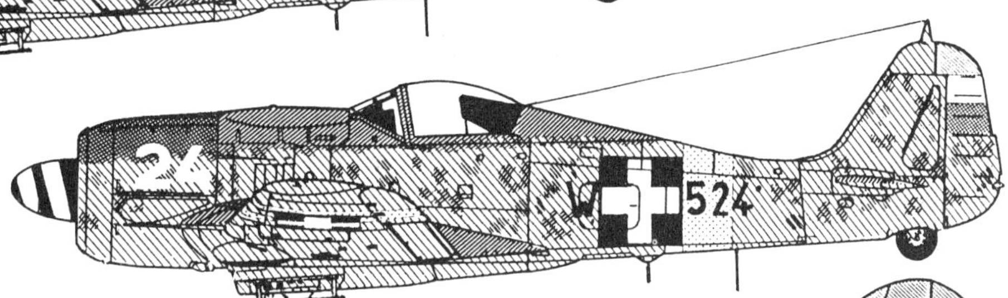

Fw 190F-8, W.5+24, of the *102/1. Vadászbombázószázad* in the Trans-Danubian area of Hungary, 1944. Standard Luftwaffe 74/75/76 finish, white number, black codes

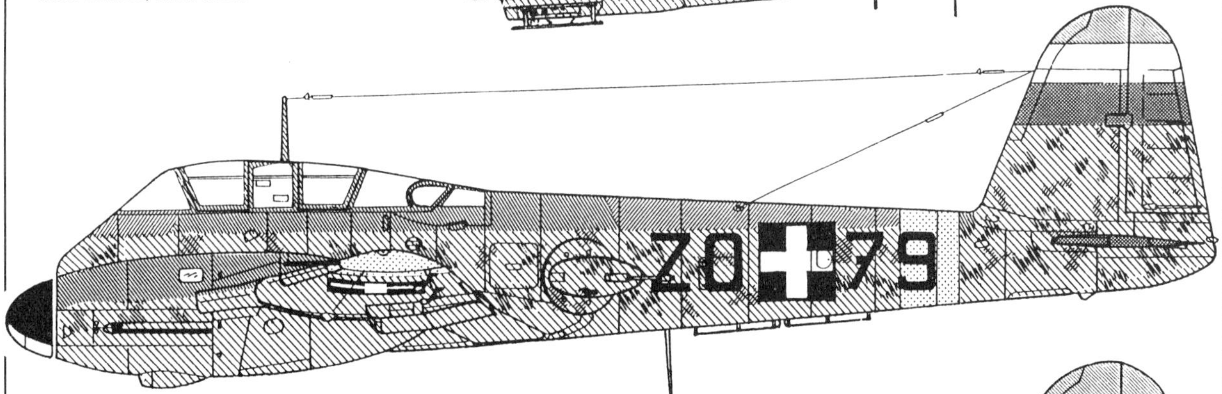

Me 210Ca-1, Z.079, *102/2. Gyorsbombázószázad,* Hajduböszörmény, 1944

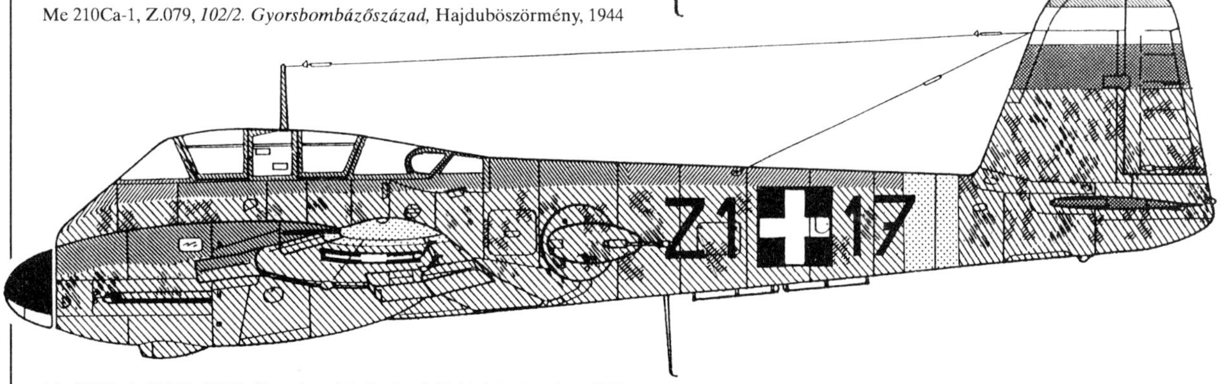

Me 210Ca-1, Z.117, *102/1. Gyorsbombázószázad,* Hajduböszörmény, 1944

1/72nd SCALE

CAMOUFLAGE AND MARKINGS

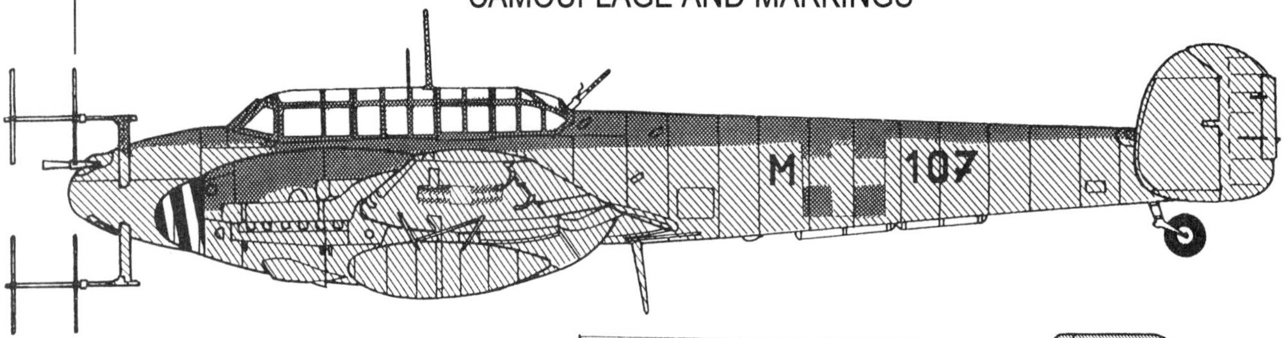

Bf 110G-4b nightfighter of the *101. Éjjeli vadászszázad*, Wiener Neustadt, spring 1945. Finish is standard Luftwaffe 75/76

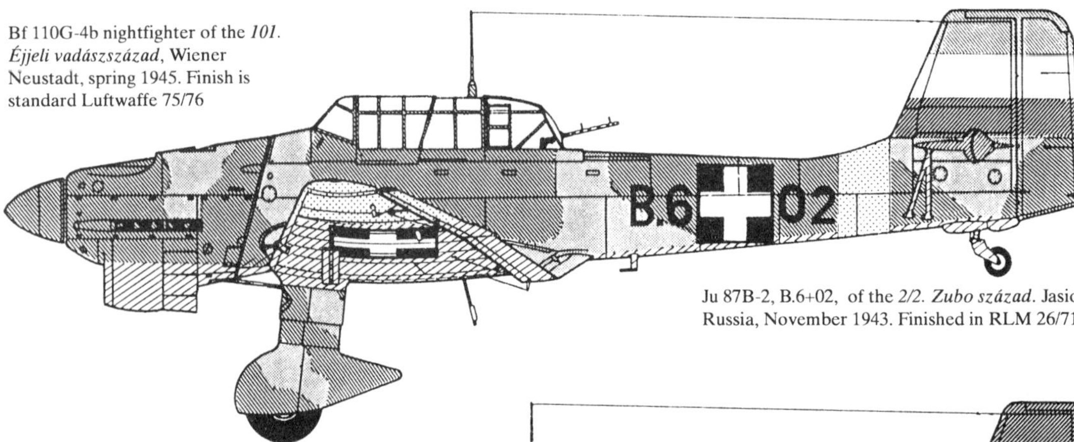

Ju 87B-2, B.6+02, of the *2/2. Zubo század*. Jasionka, Russia, November 1943. Finished in RLM 26/71/76

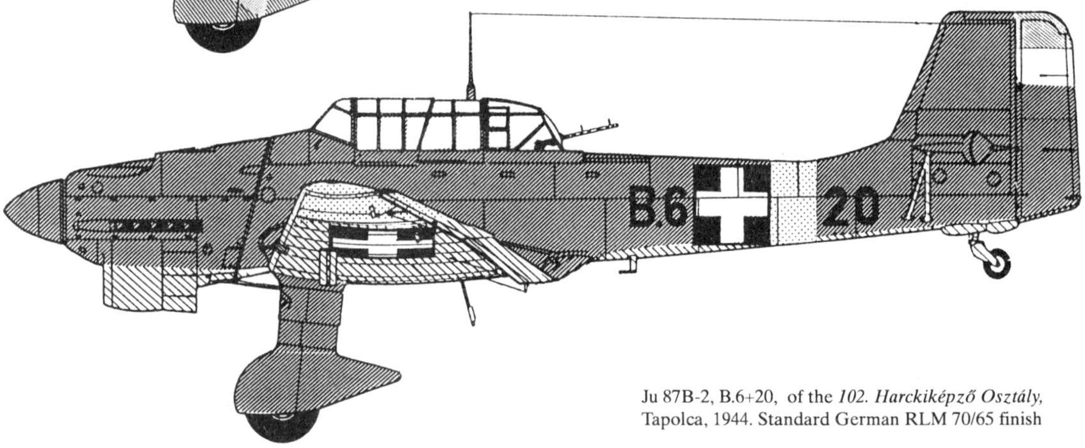

Ju 87B-2, B.6+20, of the *102. Harckiképző Osztály*, Tapolca, 1944. Standard German RLM 70/65 finish

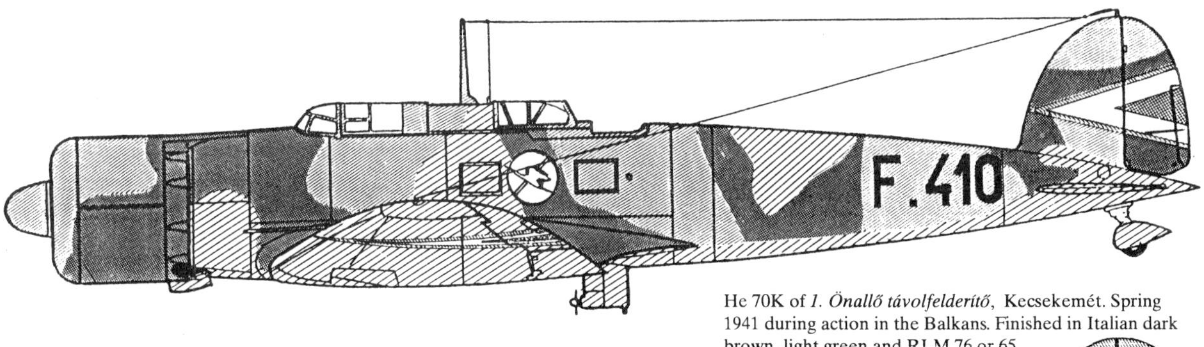

He 70K of *1. Önallő távolfelderítő*, Kecsekemét. Spring 1941 during action in the Balkans. Finished in Italian dark brown, light green and RLM 76 or 65

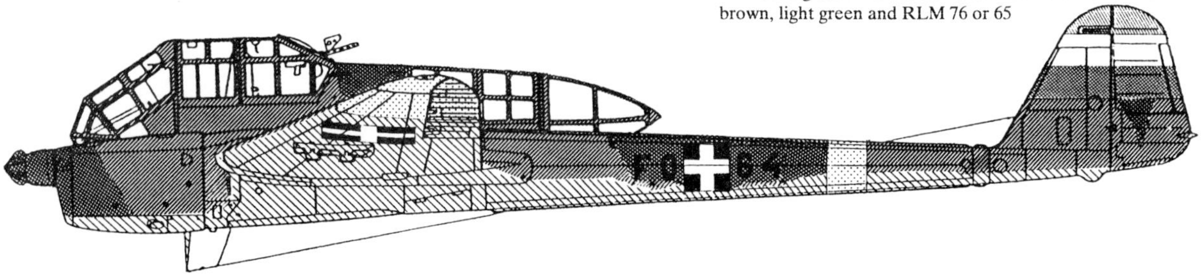

Fw 189 of *3/1. Közelfelderítő*, Kharkov, May 1943. Flown by *Zászlós* Béla Csapó. Finished RLM 70/71/65

1/72nd SCALE

CAMOUFLAGE AND MARKINGS

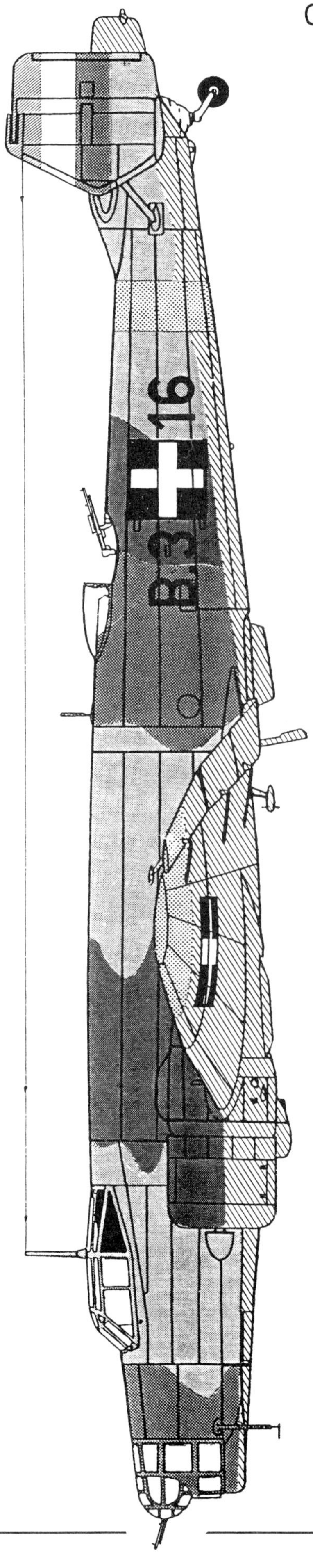

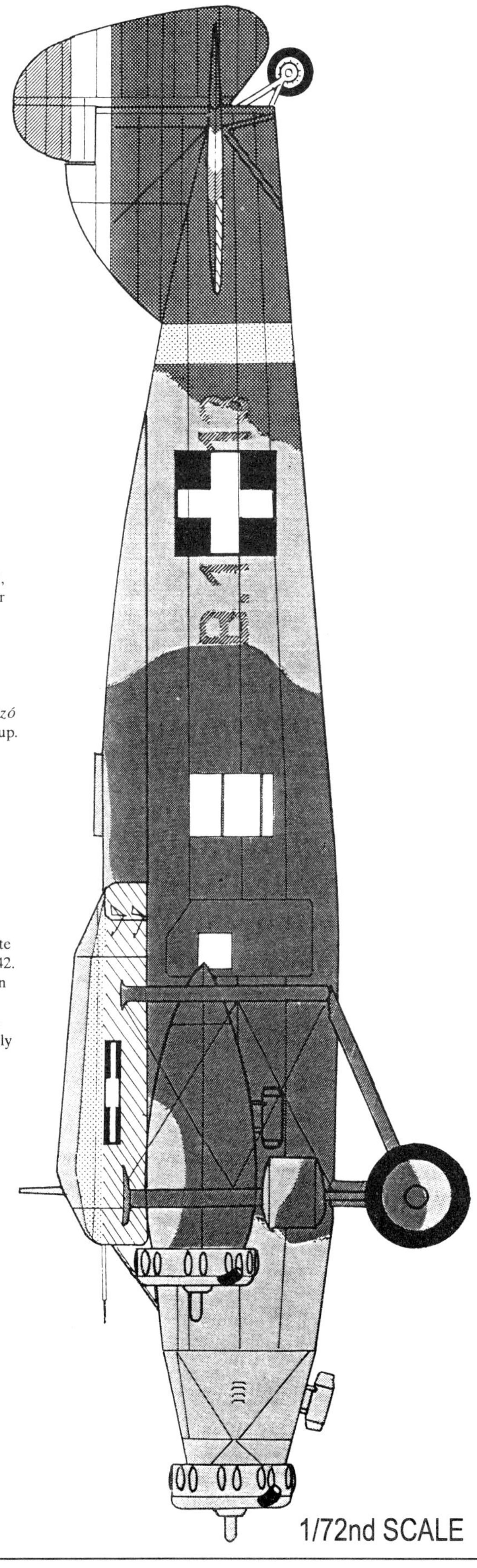

Junkers Ju 86K-2, 'B.3+16', of the 2/4. medium bomber squadron, which later became the *4. Önállő bombázószázad* – 4th Independent Bomber Squadron. Later still, the unit became the *4. Bombázó osztály* – 4th Bomber Group. Seen here in 1942 at Debrecen, the machine displays the three-tone upper surface camouflage scheme based upon a mixture of German and Italian paints

Caproni Ca 101, 'B.1+13', serving with the 1st Hungarian Army Parachute Battalion, sometime in 1942. Finish appears to be Italian red-brown and sand with light blue-grey undersides. Code numbers are probably red, or may have been overpainted

1/72nd SCALE

CAMOUFLAGE AND MARKINGS

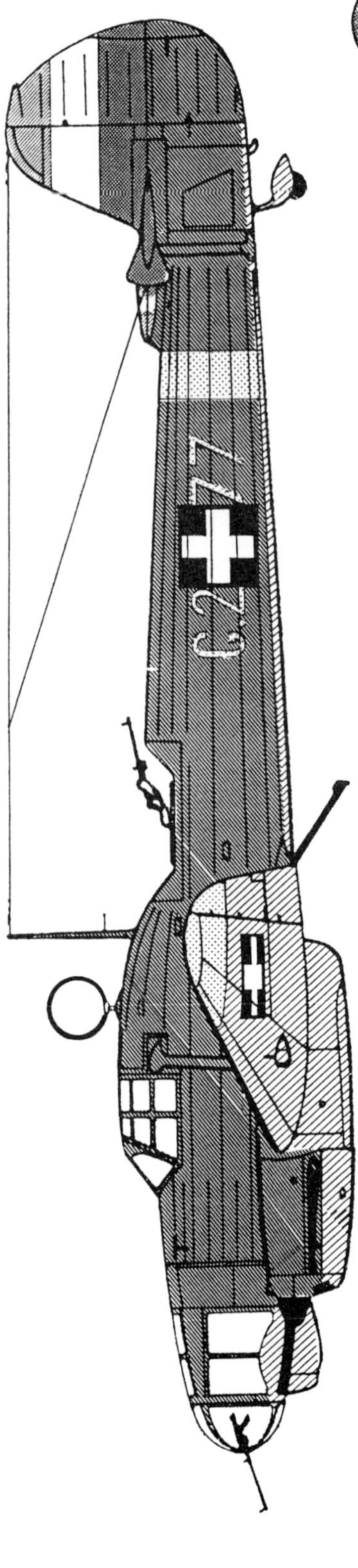

Focke-Wulf Fw 58, 'G.277', seen at the MWG factory. Finish is overall RLM 71 Dunkelgrün uppersurfaces, with RLM 65 underneath. Code is yellow

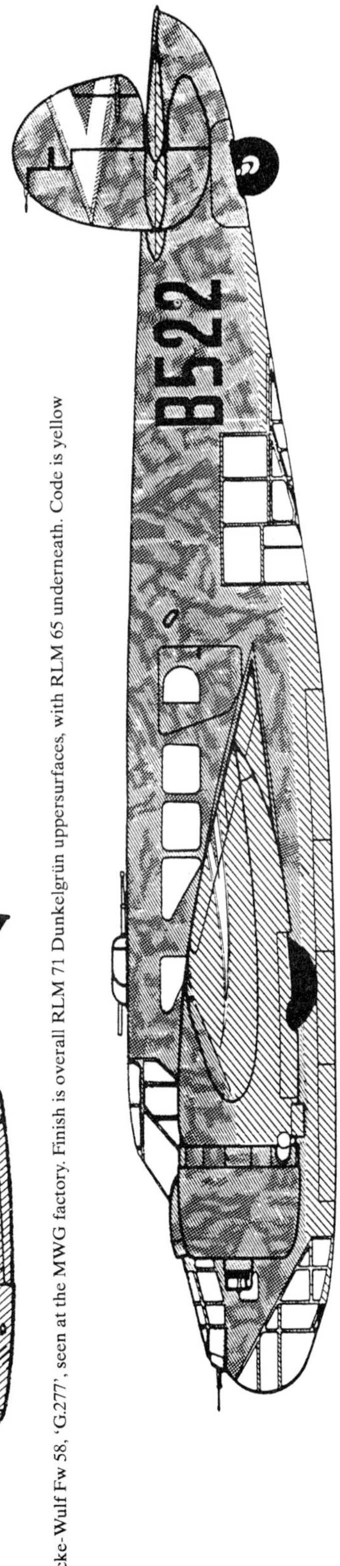

Caproni Ca 135bis, M.M. 21585, 'B.522', of the 3/6. *Bombázószázad* in spring 1941. Later the aircraft was used for radio-training. Standard Italian sand/green/light grey finish

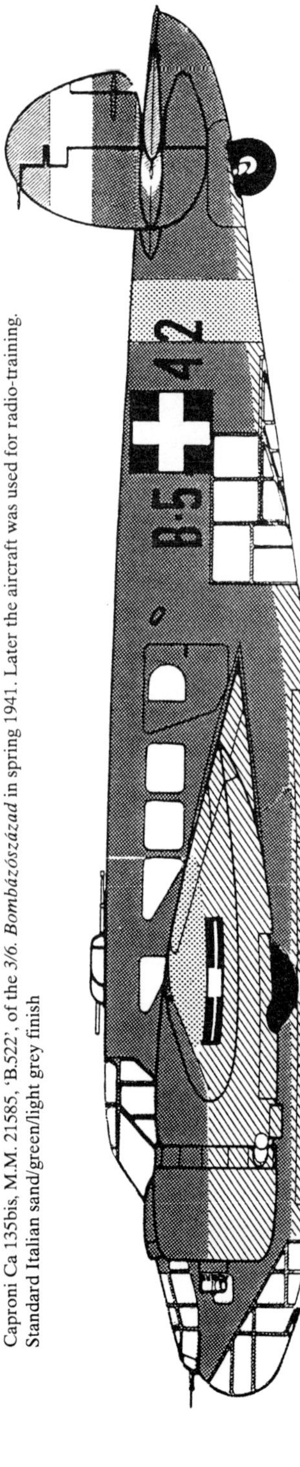

Caproni Ca 135bis, M.M. 21955, 'B.542', of the *4. Önálló Bombázóosztály* – 4th Independent Bomber Group – in summer 1942, on the Russian Front. Italian overall Italian dark green uppersurfaces, light grey below

1/72nd SCALE

CAMOUFLAGE AND MARKINGS

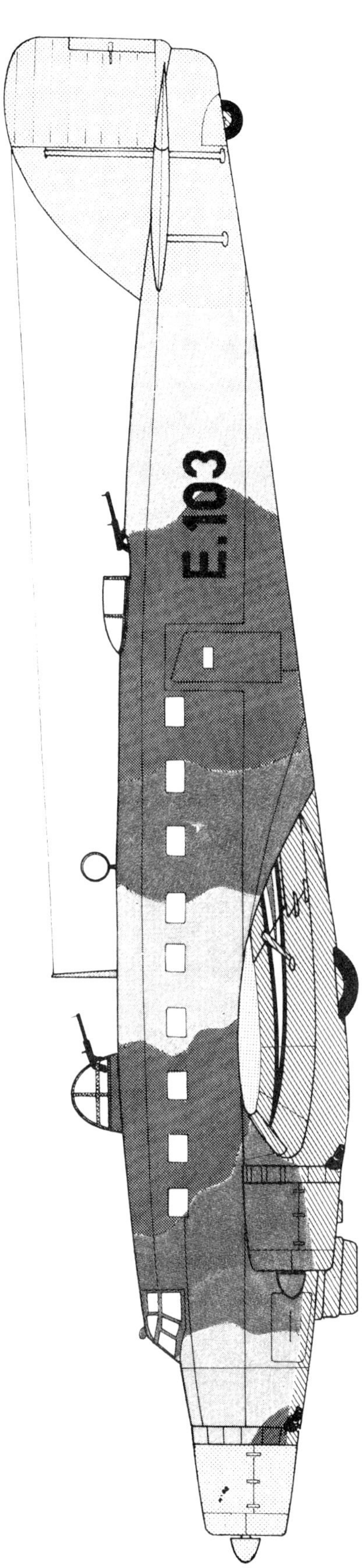

Savoia-Marchetti SM 75, 'E.103'. Flown by *Törzsőrmester* – Staff Sergeant – Sándor Szalkai, this aircraft carried the men of the Hungarian 1st Parachute Battalion from Pápa-Veszprém into action in Yugoslavia on 12th April, 1941. Uppersurface finish is a combination of bands of RLM 71 dark green and Italian sand and red-brown with light grey undersides. The centre engine cowling, wingtips and tail bear the yellow Balkan Theatre markings. Note that this illustration is not to scale

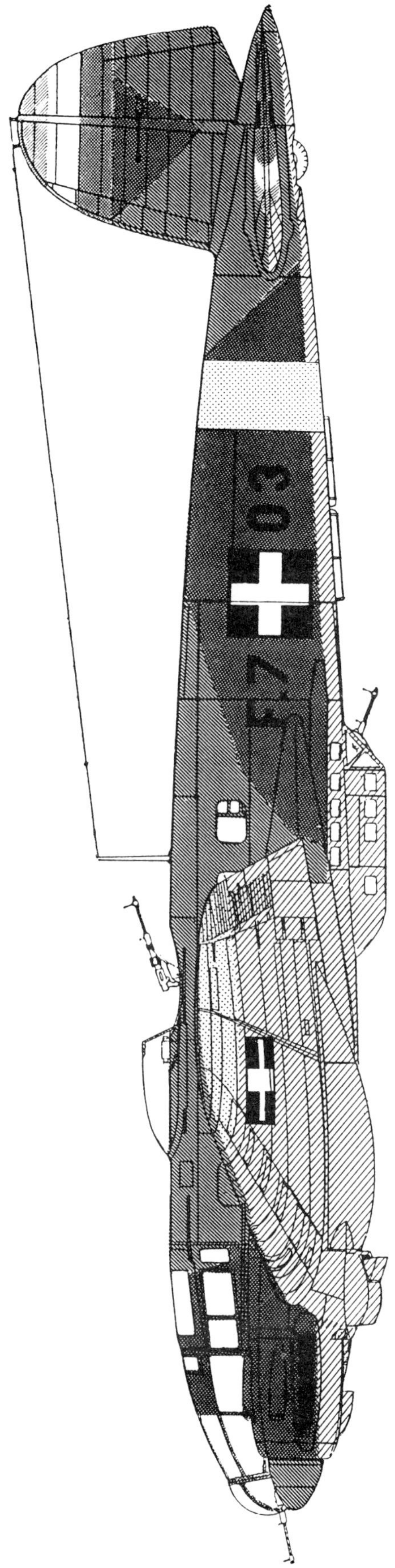

Heinkel He 111P-2, 'F.703', was a partial rebuild on the strength of the *1/1 Távolfelderítő század* – Tactical Reconnaissance Squadron – at Poltava in the Soviet Union, in January 1943. At the time the machine was piloted by *Hadnagy* – Second Lieutenant – Adolf Péterdi. Probably refinished in Luftwaffe 70/71/65 or 76 paints to a non-standard pattern in the field

1/72nd SCALE

CAMOUFLAGE AND MARKINGS

Fiat G.12 transport, probably finished in Italian dark green and light grey paints with yellow theatre markings. Date and location unknown. Note that this illustration is not to scale

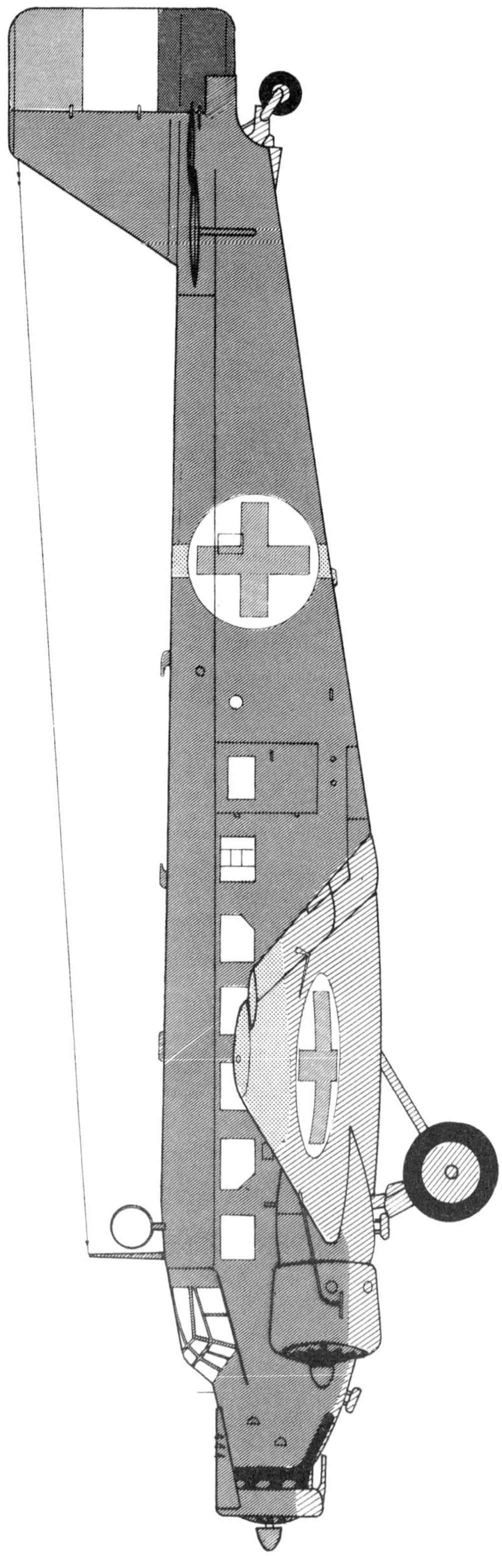

Junkers Ju 52/3m, an ex-MALERT machine, formerly HA-JEA, pressed into MKHL service as an air ambulance on the Russian Front in 1943. Dark green and light blue with red and white insignia and yellow theatre markings

1/96 SCALE

Squadron Emblems of the Magyar Királyi Honvéd Légierö 1930-1945

1 *Bagoly* - Owl - 1. Felderítoszázad, Debrecen. Fokker C.V, HD 22. 1930-1935

2 *'Meteorológiai Repülő Kirendeltseg'* - 2. Felderítoszázad, Szeged. F-2B, Fokker C.V. 1930-1935

3 *Sas* - Eagle - 3. Felderítoszázad, Kaposvár. Fokker C.V. 1930-1935

4 *Hollo* - Raven - I. Kőzelfelderíto század, Mátyásföld. He 46. 1937-1942

5 *Lucifer* - II. Kőzelfelderíto század, Székesfehérvár. He 45/46. 1936-1940

6 *Gépkocsizo* - Chauffeurs - dd Kőzelfelderíto, Budaörs, Ungvár. WM 21. 1939-1940

7 *Somogyi bicska* - Pocket Knife - III. Kőzelfelderíto század, Kaposvár. He 45/46. 1936-1940

8 *Vörös kakas* - Red Rooster - IV. Kőzelfelderíto század, Pécs. Fokker C.V. 1935-1938

9 *Paprikas boszorkány* - Paprika Witch - V. Kőzelfelderíto század, Szeged. Fokker C.V, WM 21. 1936-1940

10 *Csikos* - Horseherd - VI. Kőzelfelderíto század, Debrecen. WM 16. 1933-1940

11 *Kocsonyás béka* - lit. Jelly Frog - VII. Kőzelfelderíto század, Miskolc. WM 21. 1935-1940

12 *Sólyom* - Falcon - IV., later III Kőzelfelderíto század, Szeged. WM 21. 1938-1942

13 *Ludas Matyi* - a folklore character - VIII., later 4. Kiképző kőzelfelderíto század, Ungvár. WM 21. 1938-1942

14 *Táltos* - Pegasus - with 1. & 2. cavalry regiments, WM 16, WM 21. Later 3. Kőzelfelderíto osztálty, Kecskemét. Fw 189. 1939-1944

15 *Sólyomzem* - Hawkeye - X. Kőzelfelderíto század, Mobile Army Corps. WM 21. 1940-1942

16 *Mérföldes csizma* - Seven-league Boots - 1/1. Távolfelderíto század, Mátyásföld, Kecskemét, Budaörs. He 70/111, Ju 86/88/188. 1937-1945

17 *Gólya* - Stork - 1/2. Távolfelderíto század, Mátyásföld, Kecskemét. He 70. 1937-1941

18 *Palatábla* - Blackboard - 2. Felderítő iskolszázad, Székesfehérvár. He 45, Ar 96, Bü 131. 1937-1942

19 *Futó* - Runner - 1. Futárszázad. Ar 96, Fi 156, Fw 58. 1941-1945

20 *Hírnök* - Messenger - 102. Hadsereg-futárszázad. Fi 156, UT-2, Fw 58. 1941-1945

21 *Szállitó* - Carrier - 102/2 MALERT század, Budaörs, Veszprém. 1941-1945

22 *Ernyős* - Umbrella - 2. Ejtőernyős szállitószázad, Pápa. SM 75. 1939-1942

23 *Gavallér* - Gentlemen - 1. Ejtőernyős szállitószázad, Pápa. Ca 101. 1939-1944

24 *Boszorkány* - Witch - 2/1. Bombázószázad. Ca 101. Later 4. Bombázóosztály, Szombathely, Debrecen. Ca 135, Ju 88. 1933-1945

25 *Fekete Macska* - Black Cat - 2/2. Bombázószázad, Szombathely. Ca 101. 1933-1936

26 *Buzogány* - Winged Mace - 2/3. Bombázószázad, Szombathely. Ju 86. 1937-1940

27 *Vörös Ördög* - Red Devil - 2/4. Könnyűbombázó század, Szombathely. Ju 86, WM 16, Ca 310. 1935-1942

28 *Hálál* - Death - 2/5. Bombázószázad, Szombathely. Do 23, Ju 86. 1935-1942

29 *Isten nyila* - Arrow of God - 3/1., later 4/1. Bombázó század, Tapolca. Ju 86. 1937-1942

30 *Isten Kardja* - Sword of God - 3/2., later 4/2. Bombázó század, Tapolca. Ju 86. 1937-1942

31 *Sarga vihar* - Yellow Storm - 3/3. Bombázó század, Veszprém. Ju 86. 1937-1940

32 *Sárkány* - Dragon - 3/4. Bombázó század, Pápa. Ju 86, Ca 135, Ca 310. 1937-1939

33 *Bomba Matyi* - Tom Thumb - 3/5. Bombázó század, Pápa. Ju 86. 1937-1940

34 *Uz Bence* - Woodsman - 3/6. Bombázó század, Debrecen. Ca 135. 1940-1941

35 *Vezesd a fiatalt* - Young leaders - 1. Bombázó iskolaszázad, Szombathely. Ju 86, Ca 135. 1939-1944

36 *Sas* - Eagle - 102/1. Gyorsbombázó század, Hajdúböszörmény. Me 210Ca-1. 1944-1945

37 *Tigris* - Tiger - 102/2. Gyorsbombázó század, Hajdúböszörmény, Várpalota, Tes. Me 210Ca-1. 1944-1945

38 *Villám* - Lightning - 102/3 RKI gyorsbombázó század, Ferihegy, Várpalota. Me 210Ca-1. 1943-1945

39 *Pávián* - Baboon, aka *Kókuszdió* - Coconut - 102/1. Zuhanóbombázó század, Kolozsvár. Ju 87 1943-1944. Later 102/1. Vadaszbombázó század. Fw 190. October 1944

40 *Puli* - a dog's name - 102. Csatarepülo osztály, Páli, Börgönd. Fw 190F-8. 1945

41 *Turul* - a mythical bird - 1. Vadászcsoport, Székesfehérvár. Fokker D.VII, D.XVI. 1930-1934

42 *Ijász* - Archer - 1/1. Vadászszázad, Szombathely, Börgönd. Fiat CR 20, 32. 1935-1941

43 *Ludas Matyi* - a folklore character - 1/2. Vadászszázad, Börgönd. CR 32. 1936-1941. Héja. 1941-

44 *Puma* - 1/3., later 2/2. Vadászszázad, Nyiregyháza, Veszprém. Fiat CR 32, Héja, Bf 109. 1936-1945

45 *Kör Asz* - Ace of Hearts - 1/6., later 1/3. Vadászszázad, Kecskemét, Mátyásföld. CR 32, CR 42. 1936-1942

46 *Dongó* - Wasp - 1/5. Vadászszázad, later 1/1. Vadászszázad, Kecskemét, Szolnok. CR 32, Héja. 1936-1944

47 *Teve* - Camel - 1/4. Vadászszázad, Kecskemét, Mátyásföld. CR 32. 1936-1940

48 *Szent György* - Saint George - 1/4. Vadászszázad, Mátyásföld. CR 42. 1940-1943

49 *Mackó* - Teddy Bear - 2/1. Vadászszázad, Nyíregyháza. CR 32. 1938-1940

50 *Nyil* - Arrow - 2/4. Vadászszázad, Kolozsvár. CR 42. 1940-1943

51 *Ricsi* - Dick - 2/3. Vadászszázad, Kolozsvár. CR 42. 1940-1943

52 *Bagoly* - Owl - 5/1., later 102/1. Kísérleti éjjeli vadászszázad, Csorna. Fw 58, Héja, Me 210Ca-1, Bf 110. 1944-1945

53 *Keresztespók* - Spider Knight - 2/4. Vadászszázad, Kolozsvár. Héja. 1941-1944

54 *Ricsi* - Dick - 102. Vadászosztály, later 101/8. Század, Kolozsvár. Bf 109G-10. 1944-1945

55 *Káró Ász* - Ace of Diamonds - 101/7. Vadászszázad. Bf 109G. 1944-1945

56 *Béka* - Frog - 101/9. Vadászszázad, Veszprém. Bf 109G-10. 1945

Hungarian Military Aircraft Codes 1919-1945

Training Types

Air Ministry Type Number	Type	Function	Military Markings	Number Used	Dates Used
76	Brandenburg B.I	Primary Trainer	H-J.30	15	1919-1921
001	Oravecz-Szebeny	Primary Trainer	H-MAAA, C	2	1923-1925
011	Oravecz I	Primary Trainer	H-MAAB	1	1923-1926
021	Oravecz II	Primary Trainer	H-MAAG, -J	4	1923-1925
051	Lóczy *'Brandi'* (KLB)	Trainer	H-MAHA, -AHE	5	1926-1932
201	Bristol School	Primary Trainer	A, -E	5	1926-1930
301	Udet U 12 *'Flamingo'*	Primary Trainer	a, -e	5	1926-1930
101	Lóczy *'Hungaria'*	Primary Trainer	K.002-	80	1927-1932
351	WM-Udet U-12b	Primary Trainer	K.101-	30	1928-1932
12.01	WM-Heinkel HD 22	Trainer	K.201-	30	1927-1933
122.01	REGJÜ-Fokker C.V.D/K	Basic Trainer	K.301	15	1932-1935
501	Bücker Bü 131A/B	Primary Trainer	I.151, -256	105	1936-1938
401	Junkers A 50 (Floatplane)	Primary Trainer	HA-JAA, -C	3	1930-1938
42.01	Junkers F 13	Observer Trainer	HA-JAD	1	1932-1939
	Varga *'Káplár'*	Primary Trainer	I.301-	5	1941-1945
	Fábián *'Levente'* I	Primary Trainer	I.351-	2	1941-1945
	Bücker Bü 131D-2	PT and Courier	I.401-	150	1939-1945
	Fábián *'Levente'* II	PT and Courier	I.551-	80	1942-1945
	Bücker Bü 181	PT and Courier	I.701-	23	1942-1945
	Fábián *'Honvéd'*	Twin-engine Tr	H.001, -004	4	1942-1945
14.01	Focke-Wulf Fw 56	Fighter Trainer	G.121-	32	1937-1944
411	Breda 25	Primary Trainer	G.161, -163	3	1939-1942
13.51	Romeo Ro 41/41B	Fighter Trainer	G.181, -188	8	1939-1944
17.01	Dornier Do 23	Bomber Trainer	G.201, -203	3	1937-1944
53.64	Junkers Ju 86K	Bomber Trainer	G.211, -213	3	1937-1945
15.01	Focke-Wulf Fw 58K-7/8	Trainer	G.221-	84	1937-1945
421	WM 10/13	Primary Trainer	I.111, -119	9	1933-1944
13.01	Heinkel He 45C	Observation Tr	G.301, -306	6	1937-1944
	Arado Ar 96A	Advanced Trainer	G.401, -465	65	1939-1945
	Arado Ar 96B	Gunnery Trainer	G.451- & G.901-	45	1942-1945
	Nardi FN 305 & 315	Advanced Trainer	G.601, -652	52	1941-1944
	RWD 8	PT and Tourer	I.281, -282	2	1940-1944
	Potez 63.11	Destroyer Trainer	G.701-	40	1944-1945
	NA-16 Harvard*	Fighter Trainer	G.751-	3	1943-1945
	Rubik *'Tücsök'*	Glider Trainer	101-	5	1942-1944
	Rubik *'Vöcsök'*	Glider Trainer	201-	5	1942-1944
	Rubik *'Pilis'*	Glider Trainer	301-	5	1942-1944
	Rubik *'Kevély'*	Glider Trainer	401-	5	1942-1944
	Műegyetem M-22	Glider Pilot Tr	406-	2	1942-1944
	Rubik *'Móka'*	Glider Pilot Tr	408-	2	1944-1945
	Rubik *'Cimbora'*	Glider Pilot Tr	501-	10	1942-1944
	DFS 230	Glider School	601-	3	1944-1945
	Gotha Go 242	Transport Gldr Tr	701-	15	1944-1945
	DFS *'Kranich'*	Glider Trainer	801-	3	1943-1945

*These are most likely to have been ex-French North American NA-57 fixed-undercarriage versions of the type

Hungarian Military Aircraft Codes 1919-1945

Reconnaissance, Observation, Liaison, Transport and Courier Types

Air Ministry Type Number	Type	Function	Military Markings	Number Used	Dates Used
161	UFAG C.I	Observation/C	H-EC.1, -3	3	1919-1921
49	Phönix C.I	Armoured Obs	H-J.101, -106	6	1919-1921
269	Brandenburg C.I	Observation	H-EB.1, -2	2	1919-1928
031	Oravecz III	Observation	H-MAAL, -AAK	2	1925-1926
10.10	Bristol F.2B	Observation	H-MAEA, -AEF	6	1922-1927
10.51	Junkers A. 20	Observation	H-MAIA	1	1926-1928
11.01	Junkers A. 35	Observation	H-MAJA	1	1927-1937
11.51	Fokker C.Vd	Observation	H-MPFA, -PFH	8	1926-1932
22.02	WM-Fokker C.Vd	Observation	F.101, -150	50	1928-1942*
121.01	WM 16 *'Budapest'*	Observation	F.151, -159	9	1933-1939
22.01	Fokker C.Ve	Observation	H-MEJA, -EJW	2	1927-1932
121.10	WM 21A & B *'Sólyom'*	Obs/Tac R	F.201, -299 }	128	1938-1941
121.11	WM 21A & B *'Sólyom'*	Obs/Tac R	F.601, -629 }		
131.01	Heinkel He 46E-2	Obs/Tac R	F.301, -336	36	1937-1944
24.01	Heinkel He 70K/He 170	Long Distance R	F.401, -418	18	1937-1942
25.01	Romeo Ro 37/bis	Obs/Tac R	F.501, -514	14	1939-1941
26.01	Junkers Ju 86K-2	Long Distance R	G.312, -313	2	1939-1944
	Heinkel He 111P-2/P-4	Long Distance R	F.701, -706	6	1940-1943
	Heinkel He 111P-2/P-4	Long Distance R	F.710	1	1943
	Dornier Do 215B-4	Long Distance R	F.707, -809	10	1942-1944
	Messerschmitt Me 210/Da	Obs/Tac R	Z.006, -011	6	1943-1945
	Junkers Ju 88D-1/3/5	Long Distance R	F.901, -928	28	1942-1944
	Messerschmitt Bf 109G-8	F/Tac R	F.801, -812	12	1944-1945
	Focke-Wulf Fw 189A-1/2	Tac R	F.051, -071	20	1943-1944
	Focke-Wulf Fw 189	Mapping	I.001	1	1944-1945
40.01	Fokker F.VII	Ambulance	HA-FKA, -FKB	2	1928-1938
40.51	Fokker F.VIII	Ambulance	HA-FNC, -FND	2	1930-1938
42.01	Junkers F.13/efg	Transport/Trainer	HA-JAD (D-6)	1	1930-1937
40.21	Fokker F.XI *'Universal'*	Photographic	HA-FUA, -FUB	2	1930-1938
41.01	Caproni Ca 97	Mapping	HA-ALD, -ALG	4	1930-1938
16.01	Messerschmitt Bf 108B	Courier/Spotter	G.351, -357	7	1937-1945
15.06	Focke-Wulf Fw 58C	Courier	HA-XBG	1	1937-1945
	Focke-Wulf Fw 58E-1	Weather R	G.235, -236	2	1939-1945
	Savoia-Marchetti S.M. 75	Transport	E.101, -105	5	1938-1945
	Fiat G.12	Transport	S.151, -162	18	1942-1945
	Junkers Ju 86Ga-1	Courier	G.308, -309	2	1943-1945
	Junkers Ju 52/3m g14	Transport (PIRT)	S.201, -223	23	1943-1945
	Dornier Do 17Ka-3	Long Distance R	J.101	1	1941-1944
	Fieseler Fi 156C/D	Courier	R.101, -136	36	1942-1945
	Műegyetemi M-24	Courier	I.284, -285, I.851, -852	4	1940-1942
	Arado Ar 79B	Courier	I.001, -016	16	1944-1945
	EMESE EM-29	Courier	M.001, -025	25	1942-1945
	Caudron C 600 *'Aiglon'*	Courier	I.901, -906	6	1943-1945
	Caudron C 630 *'Simoun'*	Courier	S.301, -306	6	1943-1945
	Siebel Si 204D	Courier	S.400, -405	5	1944-1945
	Focke-Wulf Fw 58K-1/2	Transport	S.001, -003	3	1943-1945
10.01	Junkers Ju 52/3mg 3-9	Passenger Trspt	U.861, -867, later S.101-107	7	1941-1945

* F.102 built as single-seat *'Furricane'* with enclosed cockpit

Hungarian Military Aircraft Codes 1919-1945

Bomber, Dive Bomber and Attack Types

Air Ministry Type Number	Type	Function	Military Markings	Number Used	Dates Used
161.	UFAG C.I	Bomber	H.EC.4-	2	1919-1922
369.	Brandenburg C.I	Bomber	H.EB-3/H-MADA	1	1919-1926
21.01	Fokker C.Ve	Light Bomber	H-MEJW, A & B	3	1928-1933
50.01	Fiat BR 3	Bomber	HA-AAF	1	1932-1935
41.21	Caproni Ca 101/Bi	Bomber	H-MAMA, -B	2	1930-1933
51.01	Caproni Ca 101/3m	Bomber	B.101, -115	15	1931-1944
52.01	WM 16 *'Budapest'*	Light Bomber	B.201, -209	9	1933-1942
53.01	Junkers Ju 86K-2	Bomber	B.301, -363	63	1937-1945
54.01	Caproni Ca 310 *'Libeccio'*	Light Bomber	B.401, -435	35	1938-1940
54.51	Fiat BR 20	Heavy Bomber	B.451, -452	2	1938-1939
	Caproni Ca 135bis	Heavy Bomber	B.501, -568	68	1940-1944
	Junkers Ju 87B-2	Dive Bomber	B.601, -614	10	1941-1943
	Junkers Ju 87A	Dive Bomber	B.603, -606	4	1941-1944
	Junkers Ju 87D-1/D-3	Dive Bomber	B.617-	30	1943-1944
	Junkers Ju 87D-5	Dive Bomber	B.701 (Later ZB.01-)	18-20	1944-1945
	Junkers Ju 88A-4/A-5	Bomber	B.401-	51	1943-1944
	Junkers Ju 88A-14/C-6	Bomber/ Heavy Fighter	B.121-	32	1944-1945
	MÁVAG Re 2000 *'Héja'*	Dive Bomber	V.560 & V.670	2	1943-1945
	Focke-Wulf Fw 58C/B-2	Night Attack	K.330 (20)	17	1944-1945
131.20	Heinkel He 46E-2 Un	Night Harassing	F.320-	12	1943-1945*
	Fiat CR 42bis	Light Attack	V.260-	12	1944-1945**
	Heinkel He 111H-6	Night Bomber	B.801-	12	1944-1945

* 12 aircraft remained from the first series
** 12 aircraft remained from the first 72 delivered

Formation Terms of the Magyar Királyi Honvéd Légierö

The Hungarian air force was subject to many pressures, both from combat and the general shortage of resources which bedevilled it throughout its existence. As a result, organisation and units were subject to frequent changes. The terms given to different formations below are those the reader will encounter most frequently, but it should be remembered that the number of aircraft involved was subject to great variation. English equivalents are approximations only. *Ezred* and *Század* were indicated by Arab numerals. *Osztály* was shown in Roman numerals.

Géppár	Pair	2 aircraft
Raj	Flight	3 aircraft
Század	Squadron	9-12 aircraft
Osztály	Group	2-3 Század
Ezred	Regiment	2-3 Osztály
Dandar	Brigade	3-6 Század
Repülő hadosztály	Air Division	
Hadsereg	Army	
Katonai	Military	

Vadász	Fighter
Bombázo	Bomber
Zuhanóbombázo	Dive Bomber
Önalló	Independent
Tavolfelderitő	Short-range/Tactical R
Közelfelderitő	Long-range/Strategic R
Szállító	Transport
Gyarkorló	Practice/Training
Iskola	School/Primary

Hungarian Military Aircraft Codes 1919-1945

Fighter and Destroyer* Types

Air Ministry Type Number	Type	Function	Military Markings	Number Used	Dates Used
92.	Aviatik Berg D.I	Fighter	H-AA.1, -4	4	1919-1925
90.	Fokker D.VI	Fighter	H-AF.1, -2	2	1919-1920
30.01	Fokker D.VII	Fighter	H-MFOA, -FOO	4	1919-1944
30.11	Fokker D.XVI	Fighter	H-MAGS, -MAGU	2	1928-1936
30.21	Ansaldo A.C. 3	Fighter	H-MAL, -MALK	2	1929-1930
30.31	Fiat CR 20/20B	Fighter	H-MAPP, -APQ	2	1930-1934
31.01	Fiat CR 20bis	Fighter	Z-MAZA,-	16	1932-1937
30.51	Fiat CR 30	Fighter	G.101, -110	10	1936-1939
30.61	Fiat CR 30/30B	Fighter Trainer	G.111, -117	2+5	1936-1939
33.01	Fiat CR 32/32bis	Fighter	V.101, -176	76	1936-1942
133.01	Fiat CR 32ter/32quater	Fighter	V.080, -100 & V.185, -200	36	1939-1944
34.01	Heinkel He 112B-0	Fighter	V.301, -303	3	1939-1944
35.01	Fiat CR 42	Fighter	V.201, -272	72	1939-1945
	Messerschmitt Bf 109D-1	Fighter	V.601, -602	2	1941-1942
	Reggiane Re 2000 *'Falco'*	Fighter	V.401, -470	70	1941-1943
	MÁVAG Re 2000 *'Héja'-M*	Fighter	V.471, -500	196	1942-1944
			V.502, -600		
			V.601, -666		
	Messerschmitt Bf 109F-4B	Fighter Bomber	V.001, -066 (Later V.01,-66)	66	1942-1943
	Messerschmitt Bf 109G-2/4	Fighter	V.310, -394	84	1943-1944
	MWG Messerschmitt Bf 109G-2/4/6	Fighter	V.751, -765 & V.766, -899	150	1943-1944
	MWG Messerschmitt Bf 109G-10/14	Fighter	W.001, -100 & W.101, -225	205	1944-1945
			With German markings	33	1945
	Messerschmitt Bf 109E-4	Fighter	V.901-	50	1944-1945
	Messerschmitt Bf 110E-4 Messerschmitt Bf 110G-2	Destroyer Destroyer	V.951, -956	6	1944-1945
	Focke-Wulf Fw 190F-8	Fighter Bomber	W.501, -572	72	1944-1945
			(then white 37-72)		1945
	Messerschmitt Me 210Ca-1	Destroyer	Z.001, -160	160	1943-1945**
			German built	19	1944-1945
	Messerschmitt Bf 110G-4	Night Fighter	M.101, -112	12	1944-1945

* Destroyer = Zerstörer i.e. heavy fighter

** Amongst these were 5 experimental prototypes, 3 tactical reconnaissance versions, 3 long-range reconnaissance versions, 16 night fighter versions, 6 trainers and 128 destroyers/fast bombers.

Hungarian Military Aircraft Codes 1919-1945

Experimental Types 1930-1945

Air Ministry Type Number	Type	Function	Military Markings	Number Used	Dates Used
32.01	AVIS	Fighter	HA-ARA, -ARF	6	1930-1938
	X/H & X/G	Destroyer	X/H, X/G	2	1941-1944
	Z/G	Dive Bomber T	Z/G	1	1941-1944
	V/G	Trainer	V/G	2	1941-1944
	Me/G	Trainer	Me/G	1	1942-1944
	X/U	Transport	X/U	1	1943-1944
	RMI-6 *'Szunyog'*	Small aircraft	Sz/G	1	1942-1944
	Marton X/V	Fighter	X/V	1	1943-1944
	WM-23 *'Ezűstnyil'*	2-seat Fighter	V.501	1	1940-1942
	Rubik R-14 *'Pinty'*	Trainer	HA-MAL	1	1942-1945
	Fábián-Burodics *'Mokány'*	Trainer	MG.01	1	1943-1944
	Rubik R-21	Transport Glider	901-	5	1942-1944
	Jancsó-Szegedi ÉM-27	Trainer	HA-NBU	1	1944

Aircraft used, but not as established types 1940-1945

Air Ministry Type Number	Type	Function	Military Markings	Number Used	Dates Used
	Curtiss P-1	Fighter	He 43	2	1930-1931
	De Havilland Gipsy Moth	Courier	I.143	1	1939-1942
	Hirtenberg HS-9	Tourer & Trainer	I.286	1	1939-1945
	Avia B-534	Fighter	G.192	3	1939-1944
	P.Z.L. P.11a	Fighter	G.191	1	1939-1944
	Praga E-39	Courier	HA-VAA	12	1939-1944
	IAR 37	Tac R	R.001	1	1940-1941
	Potez 25	Observation	J.102	1	1941-1944
	Breguet XIX	Observation	J.103	1	1941-1944
	Bristol Blenheim I	Fast Transport	J.104	1	1941-1944
	Savoia-Marchetti S.M. 79	Heavy Bomber	J.105	1	1941-1942
	Bánhidi *'Gerle'*	Tourer & Courier	I.829	1	1940-1941
	Yakovlev UT-2	Courier	0.001	1	1942-1944
	Ilyushin Il-2	Ground Attack	0.002	1	1942-1944
	Yakovlev Yak-4	Heavy Fighter	0.003	1	1942-1943
	Heinkel He 111H-2	Courier	CI+FB (German markings)	1	1944
	Henschel Hs 129B	Ground Attack	Z.501-*	4	1943

* Probably carried German markings